Other books by Brian Martin from Irene Publishing:

Nonviolence Unbound (2015)
The Controversy Manual (2014)
Whistleblowing: A Practical Guide (2013)
Backfire Manual (2012)
Doing good things better (2011)

All can be acquired from www.lulu.com/spotlight/johansen_jorgen

Ruling tactics
Methods of promoting everyday nationalism, how they serve rulers and how to oppose them

Brian Martin

Published 2017 by Irene Publishing
Sparsnäs, Sweden
http://www.irenepublishing.com/
irene.publishing@gmail.com

ISBN 978-91-88061-17-1

CONTENTS

Acknowledgements

Over many decades, numerous individuals have helped me better understand the topics covered in this book. It is truly a gift to be able to share ideas with others and be part of an ongoing conversation.

Over several years I presented extracts from work in progress to members of the high-output writing group at the University of Wollongong. For their helpful textual suggestions and feedback on content, I thank Paula Arvela, Emma Barkus, Anu Bissoonauth-Bedford, Trent Brown, Rae Campbell, Nicole Carrigan, Kathy Flynn, Xiaoping Gao, Alfie Herrero de Haro, Anneleis Humphries, Jann Karp, Michael Matteson, Anne Melano, Ben Morris, Anco Peeters, Mark Richardson, Holly Tootell, Wendy Varney, Jody Watts, Malcolm Whittaker, Angela Williams, Amy Conley Wright and Tshering Yangden.

For valuable feedback on specific chapters or points, I thank Michael Billig, Jules Boykoff, John Breuilly, Aloysia Brooks, Louise Cook-Tonkin, Giliam de Valk, Sharif Gemie, Richard Gosden, Richard Jackson, Jørgen Johansen, Jason MacLeod, Andrew Rigby, Mary Scott, Janjira Sombutpoonsiri and Steve Wright. My greatest thanks go to Tom Weber, who generously scrutinised the entire manuscript and offered invaluable advice and support.

1
Introduction

Many people love their country. They think it's the greatest. It's the place where they want to live. They defend it against criticism. They may even be willing to die or kill for it.

Some phrases indicate unwavering loyalty. A US example is "My country, right or wrong." Others phrases condemn those who are disloyal. In the US, "unAmerican" is a term of contempt, and social critics may be told, "Love it or leave it."

However, even among critics, feelings of national pride or identification are common. When it's time for the World Cup, how many soccer fans cheer for a team from a country where they have never lived nor have any family or personal connections? How many people care more about the economic prosperity of people in Bangladesh or Togo than those in the country where they live?

Thinking from the viewpoint of a country—including its people, its government and its social institutions—can generate enormous passions. This commitment could be called patriotism or nationalism, but it is broader than this—it is a way of understanding the world and one's place in it.

There's no good word to describe this sort of thinking. It might be called "countryism," except there's more involved than the country. A key part of the equation is the link between loyalty to and identification with a country and loyalty to and identification with the government and its related functions, commonly called the state. These are certainly not the same. You might love a country and hate its government. But government supporters have another agenda: they want to tie country loyalties to support for dominant social arrangements, including the government itself and, more generally, the distribution of wealth and power. This doesn't come naturally, so a lot of effort is devoted to shaping the way people think about the world. This includes thinking of the world as naturally being divided into countries ruled by governments, not questioning the distribution of wealth and power in any fundamental way, and not pushing for radical alternatives.

Why tactics are needed
In human prehistory, people lived in small bands, probably no larger than one or two hundred people. In these groups, loyalty could be vital for survival, so it is plausible that humans are predisposed to form group loyalties. In today's world, though, the groups are much larger. Instead of a hundred people, where you know everyone else and have many close personal bonds, today many countries have millions of residents. Loyalty is now to an abstraction, a group of symbols, rather than attached to individuals you interact with daily. How did the human predisposition towards group loyalty become reoriented to country-level emotional commitments?

My aim here is to illustrate some of the techniques used to build identification with dominant social institutions—including inequalities in wealth and power—as embedded within a country. My assumption is that state-centred thinking is not natural or automatic, but has to be forged and continually reinforced in relation to other commitments. By recognising and understanding the techniques involved, it may be easier to question, challenge and replace them.

In doing this, I do not assume love of country is always bad. Sometimes it serves noble purposes, as in willingness to support others in need. In many cases it is unimportant, as in choosing what clothes to wear. My concern is about country-centred thinking when it is exploited to serve damaging activities, for example constructing weapons of mass destruction or exploiting foreigners.

Scholars have analysed patriotism and nationalism, and in chapter 3 I discuss the work of a few of them. My aim here is more practical, namely to highlight some of the day-to-day efforts and activities that reinforce country-centrism and to suggest this is not something inherent in humans but rather one possible way loyalties can be assigned.

The next step is to point to alternative ways of assigning loyalties. Again, many have argued for alternatives. For example, rather than the United Nations, which is built around states, some globalists have supported a world parliament. Then there are individuals who try to transcend their formal citizenship and instead think and act as global citizens. Out of the multitude of alternatives, I focus on those that involve greater freedom, equality and justice.

After this, the following step is to look again at tactics, this time at tactics to counter ruling tactics and instead promote alternatives. The number of possible examples is huge, so I proceed by looking at particular arenas, for example sport and language, looking at two sorts of tactics. Firstly there are counter-tactics, challenging ruling tactics, and secondly there are tactics to promote alternatives.

My main aim is to show an *approach* to analysing tactics. After you start noticing the use of everyday methods to promote patriotism or to encourage thinking of the world as a set of countries, you are in a better position to recognise alternatives and to understand strategies for resistance and building alternatives. Whether to join these efforts of course is a matter of choice.

Chapter 2 describes research on "moral foundations" that is useful for putting ruling tactics in context. Chapter 3 discusses ideas from a few key writings about nationalism. In subsequent chapters, I canvass various areas where ruling tactics can be observed in everyday life. These chapters can be read independently. As will be seen, the patterns are similar, though the arenas involved are quite different. My aim is less to provide a comprehensive case than to show how an analysis of tactics can proceed. Other possible areas for analysis include disability, disease, employment, environment, gender, history and technology.

Chapters 4 to 13 each begin with a general discussion of the issues, followed by an examination of specific tactics, using *some* of the following categories.

System-support tactics
1. Exposure (of positives); attention
2. Valuing
3. Positive interpretation
4. Endorsement
5. Rewards

System-support tactics: opposing challenges and alternatives
1. Cover-up
2. Devaluing
3. Negative interpretation
4. Discrediting endorsements
5. Intimidation

Opposing system-support tactics
1. Exposure (of negatives)
2. Devaluing
3. Negative interpretation
4. Discrediting endorsements
5. Refusing rewards

Promoting alternatives
1. Exposure
2. Valuing
3. Positive interpretation
4. Endorsement
5. Rewards

My interest is not just in the more ardent forms of national chauvinism but more generally in how people think of the world in terms of countries and their governments, what Michael Billig calls "banal nationalism."[1] I chose the title *Ruling Tactics* because thinking in terms of nations serves rulers. However, rulers use a host of other tactics too, hence the long descriptive subtitle.

How I got onto this topic
For many years, I've been interested in strategy for social movements, for example the environmental and peace movements. How can activists be more effective in pursuing their goals? My special interest has been in nonviolent action, including methods such as rallies, strikes, boycotts and sit-ins. Most of the effective social movements, including the anti-slavery, labour and feminist movements, have relied primarily on nonviolent methods.

When activists mount a campaign, sometimes the government uses force against campaigners, with arrests, beatings and shootings. Occasionally, government repression doesn't work: it generates huge outrage and triggers greater popular resistance. For example, in 1960 in South Africa, police shot into a crowd of protesters in the town of Sharpeville, killing about a hundred of them. Journalists were present and photos were taken. The Sharpeville

---

1 See chapter 3.

massacre undermined the South African government's credibility internationally.[2]

However, instances in which government repression is counterproductive are rare. I started looking at the methods used by governments to reduce outrage, and came up with five main methods: cover up the action, devalue the target, reinterpret what happened (through lying, minimising consequences, blaming others and using favourable framing), use official channels to give an appearance of justice, and intimidate or reward people involved. After the Sharpeville massacre, the South African police and government used all these methods.[3]

The next step is to look at counter-methods. These are exposing the action, validating the target, interpreting the events as an injustice, avoiding official channels and instead mobilising support, and resisting intimidation and rewards.

Before long I was looking at all sorts of issues in terms of tactics, including bullying at work, sexual harassment, censorship, torture and genocide.[4] Tactics are just methods, and to refer to tactics doesn't necessarily imply that people are sitting around plotting what they are going to do. Most tactics are instinctive in the sense that

---

2 The authoritative source is Philip Frankel, *An Ordinary Atrocity: Sharpeville and its Massacre* (New Haven, CT: Yale University Press, 2001).

3 Brian Martin, *Justice Ignited: The Dynamics of Backfire* (Lanham, MD: Rowman & Littlefield, 2007). Chapter 2 is on the Sharpeville massacre.

4 "Backfire materials," http://www.bmartin.cc/pubs/backfire.html

people use them without carefully considering options for achieving goals, though subsequently they often think up rationalisations for their actions. At Sharpeville, just after the police had shot and killed protesters, they removed some of the bodies whose injuries revealed the use of dum-dum bullets, banned internationally. This served to hide evidence but, like the shooting, the removal of bodies was unplanned, not the result of a thoughtful consideration of alternatives.

At some point, I started thinking about tactics in relation to patriotism and nationalism. As discussed in chapter 2, patriotism and nationalism are not natural. In fact, quite a few people are critical of them. What maintains thinking in terms of nations and maintains loyalties to particular nations? I decided to apply my tactics framework to the topic, and this morphed into the categories listed above. I then picked some of my favourite topics and looked for examples. I find it fascinating to see how easily thinking (including my own) can be channelled, and challenging to figure out how to think and act differently. This book is part of my journey. I hope you can see what's involved and find your own path, whatever it may be.

2
Moral foundations

What makes a person think it is good to be patriotic? To help understand the need to foster identification with a country and its institutions, it is useful to study the work of Jonathan Haidt on the foundations of morality.[1] Haidt is a psychologist who wants to understand why people make commitments to particular religions and political parties, among other things. Here I outline some of Haidt's ideas, noting their relevance to understanding why efforts are needed to encourage citizens to identify with their country.

The rider and the elephant
As a preliminary, Haidt presents the view that each of us has two minds.[2] One mind is intuitive, automatic and high capacity. If you see a rock approaching your head, it is

---

1 Jonathan Haidt, *The Righteous Mind: Why Good People Are Divided by Politics and Religion* (New York: Pantheon, 2012). Haidt and his collaborators have written many detailed technical articles.

2 This view is standard among psychologists. See, for example, Jonathan St B. T. Evans, *Thinking Twice: Two Minds in One Brain* (Oxford: Oxford University Press, 2010); Daniel Kahneman, *Thinking, Fast and Slow* (New York: Farrar, Straus and Giroux, 2011); Timothy D. Wilson, *Strangers to Ourselves: Discovering the Adaptive Unconscious* (Cambridge, MA: Harvard University Press, 2004).

valuable to duck without pausing to calculate the trajectory of the rock or indeed determine whether it actually is a rock rather than an illusion. In early human evolution, such an automatic system improved the odds of surviving. Responding quickly and automatically to suspicious sounds could enable escape from a predator, and was advantageous even if most such sounds were false alarms.

In the modern world, the intuitive mind still rules much of people's behaviour. A soldier learns to respond quickly to the sound of gunfire and, after returning to civilian life, may hit the ground at the sound of a car backfiring.

The other major component of the mind is slow, methodical and low-capacity: it takes more effort. It is the part of the mind commonly thought of as rational. It weighs up evidence, considers options and draws conclusions, and then may assess them on the basis of new evidence. Scientific research, in its ideal form, relies entirely on this sort of rational evaluation.

Haidt calls the intuitive mind the elephant and the rational mind the rider. In Haidt's metaphor, the rider sits on top of the elephant, perhaps trying to steer it but in most cases actually being at the mercy of the elephant's whims. The elephant is too strong and independent for the rider to control it except in carefully constructed circumstances. What often happens is that the elephant goes in a direction and the rider simply follows: the intuitive mind reaches a conclusion and the rational mind then figures out reasons to justify this conclusion.

Haidt provides some illuminating examples that are, by design, uncomfortable or even repellent for some people asked to consider them. One is a hypothetical situation of a brother and sister who are travelling together and decide to have sex with each other just to see what it is like. They do it just once, each of them using birth control. They enjoy it but decide not to do it again. The first question: is this right or wrong? The second question: why? Many people immediately react by saying it's wrong. That is their elephant speaking. But they find it challenging to explain why. Some say it is because of the possibility of conceiving a child with genetic defects, ignoring the information about birth control. The rider casts about for a plausible justification of the elephant's choice, but in this case gets stuck.

With other issues, the rider has more options. Consider the issue of drugs such as heroin and cocaine. Many people react intuitively to say they should be illegal. In a debate with a proponent of harm-minimisation, who recommends decriminalisation or legalisation, people might say the dangers are too great, that enforcement needs to be stronger and any of a host of other reasons. But they seldom argue for making alcohol or nicotine illegal. The same applies to those on the other side: they too can come up with many reasons to justify their views. Seldom does someone say, "I don't really know which drugs should be illegal, if any, because I haven't studied the issue in enough detail."

The elephant usually prevails even when the rider is more sophisticated. People with greater intelligence may simply be better at developing clever arguments to justify positions they have taken on intuitive grounds. Intelligence is not a guarantee against bias and prejudice.

An effective counter to misguided views is other people who point out shortcomings. This is most apparent in scientific research. Many scientists, including leading scientists, are strongly committed to their viewpoints, so much so that new evidence will not budge them: they simply come up with ingenious reasons why the contrary evidence is wrong or irrelevant and why their position is still viable. This was shown in a classic study of 40 scientists involved in studying rocks from the moon. Following the first voyages to the moon and return of moon rocks to earth, there was lots of new evidence that could be used to adjudicate between different theories about the origin and nature of the moon. However, key scientists who were advocates of different theories, and who were considered by their peers as especially out-standing in the field, were highly resistant to changing their views. This study showed that commitment plays a crucial role in science and that the idea that scientists seek to falsify their theories is not the way science operates in practice.[3]

So for scientists, the rider sometimes serves to justify a gut reaction, especially commitment to a viewpoint on which they have built careers and reputations. What make a difference, eventually, are other scientists. Those without prior commitments or who are more open to

---

3 Ian I. Mitroff, *The Subjective Side of Science: A Philosophical Inquiry into the Psychology of the Apollo Moon Scientists* (Amsterdam: Elsevier, 1974). See also Michael J. Mahoney, *Scientist as Subject: The Psychological Imperative* (Cambridge, MA: Ballinger, 1976).

evidence may adopt different views. More importantly, scientists with contrary views will point out flaws in evidence and logic. The rider-elephant combination may not change direction on its own, but other rider-elephants, going in different directions, sometimes can have an impact.

If this sort of commitment is common in science, with all its systems for peer review and emphasis on rigour, it is even more likely to prevail in politics. After someone develops loyalty to a political party, for example, they may stick with it tenaciously. The elephant has formed a preference and the rider will try to figure out a justification.

The six foundations
Haidt argues that people's moral judgements—their judgements about right and wrong—are influenced by six elements or reference points. He calls them moral foundations. They are care, fairness, liberty, loyalty, authority and sanctity.

People's behaviour and thoughts are potentially influenced by each of these foundations. Haidt says they are deeply embedded in human evolution and social interactions. However, an individual's foundations can change through various processes.

Care means caring for others. The most obvious instance is looking after children, something that most mothers seem to find instinctive. Small groups of humans that did not care for their children would have had a hard time surviving. Furthermore, caring for other adults in the group was also advantageous, because otherwise individuals might compete with or even attack each other, undermining the capacity of the group as a whole to survive.

In modern-day societies, the care foundation manifests itself in support for those who are disadvantaged, for example people with disabilities, those who are ill, people in poverty—including people in remote parts of the world. The care foundation evolved from concern about vulnerable members of one's own group, but now can be extended to people anywhere in the world, and even more broadly to animals and the natural environment.

Fairness is another important moral foundation. It can be evoked when someone else receives something they apparently don't deserve. A small child may protest when a brother or sister receives a bigger portion of ice cream. In the workplace, workers at the same level may protest if a co-worker receives special privileges, such as attractive assignments from the boss, or a higher salary for the same work.

The sense of fairness doesn't always give the same results. Some people think it is unfair that those who do no paid work receive unemployment payments—they may be called spongers or welfare parasites—whereas others think it is unfair when children inherit money and property from parents, especially when they did nothing to deserve this windfall. This suggests there are many processes involved in assigning the sense of fairness to particular situations.

Liberty is the sense or demand to be free and independent of oppressive power. It is especially pronounced among libertarians, who oppose many or even most functions of government, instead supporting private solutions, such as markets or voluntary arrangements such as charity. Even those from other parts of the political spectrum are influenced by the urge for liberty. This is seen especially among people subject to repressive governments, some of whom are resentful, even when the government functions well. Those with a strong liberty orientation would oppose a benevolent dictatorship.

Loyalty involves commitment to a group, a movement or even an abstraction. People can feel loyalty to family, friends, neighbours, clubs, co-workers, employers, sporting teams, commercial brands and countries. In warfare, soldiers may feel tremendous loyalty to their closest mates, even being willing to die for them.

Loyalty to one's country is central to patriotism. This often means supporting one's own government in any contest with others.

Loyalty is often expected of others in the same group. Those who go against expectations may be called traitors. Spies are caught in the crossfire of competing loyalties: they are patriots to those on one side and detested by the other. Few people think of spies as simply doing a job.

In human prehistory, the survival of the group was vital, and loyalty to the group was highly advantageous. This is the evolutionary basis for loyalty being a moral foundation. However, people today are loyal to groups quite unlike earlier times—sporting teams, for example, have no relevance to survival, except in a metaphorical way. Even more divergent from earlier forms of loyalty is patriotism, when the commitment is to a "community" thousands of times larger than one's personal interactions. This suggests that patriotism is not automatic or natural in any sense, but instead requires active efforts to initiate, cultivate and maintain it.

Authority is a moral foundation built around acceptance of systems of formal power, hierarchy and credibility. Many people believe that authorities should be followed, whether they are government leaders, medical experts, employers, sports coaches or heads of families. Respect for or obedience to authority helps make societies more stable. If no one accepts a boss's directives, then the boss has no power and perhaps a new method of making decisions will take over.

Much of political life involves a struggle over authority. There are struggles over positions of authority, for example military coups, elections and popular uprisings against rulers. Within organisations, there are struggles for positions of influence. Authority figures of various types, from politicians to judges, seek to exert their power, often encountering resistance from other authority figures and from those lower down.

The moral foundation of authority gives an advantage to those currently in positions of power. If someone believes that formal leaders should be respected and obeyed, this makes change more difficult. Yet many authorities need to be resisted. Repressive rulers cause much suffering.

One of the important types of authority is the law, a set of rules administered by various agencies, notably police and courts. The moral foundation of authority means that obeying the law is the default for many people. However, some laws are so unjust or harmful that breaking them might seem justified—to some people, anyway. Those who heed the authority imperative may reject any sort of law-breaking.

Moral judgements can be selective. Some challenges to authority are considered acceptable, others not. For example, in the US, questioning the views of the president might be okay or not, depending on who the president is. Authority becomes more important in some arenas. In the military, obedience to authority is a foremost value, drummed into recruits, despite lip service to a higher loyalty to other values.

In Nazi Germany, the authority foundation played a crucial role in enabling mass killing and other horrific human rights violations. The famous Milgram experiments showed that this sort of obedience to authority also was widespread in the US. The subjects of the experiment believed they were administering electric shocks to someone else; following instructions from the experimenter, many would continue even to dangerous levels.[4]

Sanctity is a moral foundation built around feelings that some things are sacred and should not be treated casually or with contempt. In the US, many patriots treat the flag as a sacred object that needs to be respected. Raising and retiring the flag is supposed to be done following specified protocols. The way it is folded is specified, and the flag should never touch the ground, which would defile it. When protesters or artists treat the flag in apparently disrespectful ways—for example burning it—this is seen as sacrilegious.

---

4 Stanley Milgram, *Obedience to Authority* (New York: Harper & Row, 1974).

Moral foundations and tactics

Haidt provides considerable evidence and many arguments in support of his classification of these six moral foundations. Most individuals are affected by all six foundations, but to different degrees. There are some patterns worthy of note. Haidt compares the role of the foundations in three political orientations in the US: libertarians, liberals and conservatives. Libertarians are opposed to most government functions and want society to be run through markets. As already noted, for them the liberty foundation is dominant.

Liberals, in contrast, are primarily influenced by three foundations: care, fairness and liberty. For them, loyalty, authority and sanctity are less influential. This helps explain why liberals are likely to support measures such as unemployment benefits, progressive taxation and foreign aid.

Conservatives, Haidt discovered, are influenced more equally by all six foundations. They are more likely than libertarians or liberals to be concerned about respecting police and the flag, for example.

Although there are systematic differences between people with different political and religious views, what is striking to me is the arbitrariness of people's moral commitments. Haidt says that the six moral foundations are the "first draft of the mind": most people have innate tendencies towards caring for children (and hence caring for others in need), and so forth through all the foundations. But the way these are played out in practice depends on circumstances.

Suppose a person has a strong tendency towards being loyal. But loyalty to what? There are many potential recipients of the feeling of commitment, support and even love: sporting teams, neighbourhoods, family members and companies, as well as governments and countries. Furthermore, there are many choices involved. Does loyalty to country mean not buying foreign goods? Does it mean not caring about government crimes? Or does it mean being especially concerned about government crimes? Does it mean supporting mining companies that are extracting and exporting the country's minerals—even if the companies are foreign owned? Or does it mean supporting calls to use the minerals within the country, or calls by environmentalists to leave the minerals in the ground and maintain a pristine environment? Loyalty has many potential attachments or recipients. To say that loyalty is a moral foundation is only the beginning of understanding how loyalty operates in practice.

My interest here is loyalty to a country or its government or people or ideals. Some people are patriotic, but many are not—indeed, there are plenty of people who are anti-patriotic. However, closely related to patriotism is something more common that can be called country-centredness, which means thinking about the world from the perspective of a particular country, usually the one where one lives or where one was born, and thinking of the world as made up of countries. News stories tell of a disaster affecting a few citizens of your country and ignore thousands dying in remote parts of the world. Stories about the economy or employment focus on local implications, not implications for elsewhere, whether Albania or Zambia.

How does patriotism and, more generally, country-based thinking develop? How is it maintained? In the following chapters, I examine some of the processes involved, looking at methods, behaviours and assumptions that foster identification with a country, then at alternative forms of action and identification and finally at strategies to move towards alternatives.

3
Nationalism

The term "nationalism" refers to support for a nation. In common parlance, a nation is a country like Albania or Zambia. However, it is useful to distinguish several things: countries, nations, states and governments.

Let's start with "country." It is easiest to think of a country as a geographical area plus everything in it, including mountains, plants and people. Argentina as a country has plains and rivers, sheep, buildings and a population of 43 million.

Next consider "government." This can refer to the political rulers within a country. Governments may include both an executive, with a president and cabinet, and a legislature. In dictatorships, there may be no legislature, or only a powerless one. In parliamentary systems, the executive—including the prime minister and cabinet—is drawn from the legislature. "Government" may also include various administrative supporters for the executive and legislature, for example heads of treasury, defence and environment departments.

Closely related to government is the state. The state includes everything officially run by or owned by the government. It includes the various departments or ministries that are headed by government figures. It includes government-run institutions such as schools, police, military, railways and so forth. People's private lives are not part of the state; only when they are at work are government employees part of the state. Private corporations are not part of the state. Independent religious bodies are not part of the state. (In a few countries, like Iran and Israel, there is a state religion.)

In simple terms, the government runs the show and the state is the government plus everything it runs.

Then there is "nation," a more challenging notion. A nation can be said to be a group of people who share a common identity. This may involve shared experience, blood ties, the same language, a religion, eating habits and various traditions. Among Native Americans, tribes like the Apache, Sioux and Cherokee are called nations: they had (and to some extent still have) shared language and culture, distinct from other tribes. In Europe, nations include the Armenians, Finns, French, Hungarians and Kurds.

The complication is that nations do not necessarily correspond to countries. Most people living in Japan today might be considered members of the Japanese nation, but there are some indigenous people, for example the Ainu from northern Japan (and eastern Russia), who are a distinct cultural group, and there are some immigrants, for example from Korea, who would be part of a different nation: the Korean nation.

Then there are nations that are spread across lots of countries. The Jewish people could be considered to be a nation; they are concentrated in Israel but millions live in other countries. People of Chinese ancestry don't all live in China: many live in Malaysia, Vietnam and other countries. (And within the country of China there are numerous other national groups).

Immigration is a complication for understanding nations. Consider an Egyptian family that immigrates to New Zealand. Are they Egyptian or Kiwis? If they remain in an Egyptian enclave and maintain Egyptian culture (religion, food, language), then they might be considered part of the Egyptian nation. But if the children grow up speaking English with a New Zealand accent, play or follow Kiwi sports, join the Anglican Church (or none at all), have they become part of the New Zealand nation? Or is New Zealand a nation at all, given its mixture of Maoris, descendants of British and other European immigrants, and new arrivals from various countries?

Reference is often made to a "nation-state." This concept assumes that a nation and a state coincide. In some cases it is nearly true, but nearly always there are some indigenous people, some immigrants and some locals who have emigrated (called expatriates).

Benedict Anderson calls nations "imagined communities," and this idea has been widely taken up.[1] A community is a group of people having something in common: they live in the same neighbourhood, eat lunch together, collect stamps or whatever. An imagined community is one in which what people have in common is not something they do, but only something in their imagination, in their minds. If you live in Brazil, you

---

1 Benedict Anderson, *Imagined Communities: Reflections on the Origin and Spread of Nationalism* (London: Verso, 1991, revised edition).

cannot possible interact with 200 million other people, including ones with different religions, ethnicities and ways of life. "Brazil," as a community, as a group of people, exists primarily in the minds of the people living in Brazil, as well as in the minds of people from other parts of the world.

Isms

Let's go from the nation to nationalism. "Nationalism" usually refers to a commitment to or identification with a nation. It can involve pride. Many people are excited when "their" national team does well in the World Cup, despite having no personal connection with any members of the team. Nationalism, at the psychological level, might involve support for or identification with political leaders, policies, climate, habits or any number of other attributes. One's own country usually is contrasted with others. Nationalism involves identification with and support for *my* country, not others. For most people, nationalism is on behalf of a single country, though it's possible to identify with Africa, the European Union or the world.

Nationalism, strangely enough, is only sometimes on behalf of a nation, at least in the sense that many scholars think of nations. If we think of Canadian nationalism, it is usually connected to the whole population, including separatists in Quebec and members of First Nations. So what should this commitment to a country be called? There's no such word as countryism. So perhaps this is where the word patriotism is useful. A patriot is a person who supports their own country, and patriotism is the commitment itself.

In many cases, patriotism is harnessed to the goals of the state or government. A patriot is prone to support policies adopted by the government in relation to other governments. This is pronounced in the case of war: patriots typically support their compatriots—citizens of their own country—against enemies. The opposite of a patriot is a traitor, someone who supports the enemy.

Patriotism has its positive side, including pride in group accomplishments and a willingness to sacrifice for the good of the whole. When people in a country are doing worthwhile things, it makes sense to support them and take pride in their achievements. But there is a darker side to patriotism: it can involve supporting crimes and abuses, including military aggression, torture and genocide. In the US, there is a saying, "My country, right or wrong." Supporting "the US"—usually meaning the government's policy in international relations, when it seems in the interests of the US people—for good causes is reasonable, but why support policies and actions that are wrong?

Patriotism becomes "blind patriotism" when people take a position simply because it is identified with their country or state, even if it involves lying, unfair dealings, theft and other crimes. This sort of patriotism is common when agents of the state are involved, including political leaders and soldiers. In the US, supporting US troops in foreign wars has become unquestionable; it is a touchstone of being patriotic. Even US opponents of the government's wars are careful not to criticise the troops, restricting themselves to criticising politicians and policies. This remains true even when the troops are involved in crimes.

In 1968, during the Vietnam war—in Vietnam called the American war—US soldiers in Charlie Company went on a rampage of killing in a village named My Lai, leaving hundreds of civilians dead, including women and children. Commanders informed about the massacre did nothing. Ron Ridenhour, hearing about what had happened, collected information and sent a powerful letter to various media and politicians, but none of them would act on it. Eventually, through the efforts of investigative journalist Seymour Hersh, the story broke, a year after the massacre. However, only one soldier, Lieutenant William Calley, was convicted of any crime, and he served minimal time in prison. Many US citizens sided with Calley. On the other hand, Hugh Thompson, who had tried to stop some of the killing and who testified about what had happened, was ostracised by other troops. In the midst of the war, many people in the US did not want to know about crimes by "their" troops. It was a classic case of "my country, right or wrong"—in this case, wrong.

Related to the concept of nationalism is what can be called statism: support for the state, sometimes glorification of the state. It is often associated with dictatorships, in which the ruler is attributed superhuman capacities. One example of statism is Nazi Germany, with Hitler the father figure who could do no wrong. The massive rallies at which Hitler spoke were rituals of worshipping the state.

Nazi Germany shows a toxic mixture of nationalism and statism. The nation in this case was associated with Aryan ethnicity and culture, as distinguished from others such as Slavs, Gypsies and Jews. After the invasion of the Soviet Union, Hitler initiated the "final solution," the extermination of Jews and other non-Aryans. This could be considered the operation of the state to enforce a particular conception of the nation, using the most drastic methods.

Historically, state elites try to harness nationalism for their own purposes. But this is complicated because nations don't map onto states in a one-on-one fashion. So what state elites usually try to do might better be said to be promoting statism and countryism.

Benedict Anderson and imagined communities
As mentioned earlier, Benedict Anderson's idea of "imagined communities" is widely cited as a way of understanding how nationalism operates. In a population of one million, it is impossible to know more than a tiny fraction of the people in a country, so the "community" exists only in the minds of the people, not in direct interactions.

Anderson's book *Imagined Communities* is a highly sophisticated treatment of the origins and spread of nationalism. He uses a highbrow writing style and assumes the reader can understand short passages in French and German. This is not bedtime reading, but it does contain many insights relevant to patriotism tactics.

Many of today's patriots refer to long traditions, often talking about a homeland that has been defended or sought for centuries. Serbians talk about the battle of Kosovo in the year 1389. However, Anderson says that any such long traditions exist only in the imagination. National identity is fairly new, something that developed beginning in the late 1700s in the Americas, adopted in

Europe in the early 1800s, and then exported to Africa and Asia through imperial conquests and by providing a model for others to follow.

Anderson notes that Europeans in the year 1500 or 1700 did not think of themselves as part of a nation. Upper class Europeans were part of a house of nobility that could stretch across several of today's countries. Peasants thought in terms of the area where they lived and worked.

Anderson, drawing on the work of other scholars as well as his own studies, attributes the origin of nationalism to developments in the Americas from roughly 1760 to 1830 involving a complex interplay of administration, printing and capitalism. Spain's colonies in the Americas were divided into administrative units. Spanish-born administrators in the Americas could move from unit to unit—for example from Chile to Mexico—and climb a career ladder with the highest rungs being in Spain, the centre of empire. But American-born administrators, called creoles, were restricted to a single unit. Nationalism provided a means of mobilising the population to throw off the restrictions imposed by Spanish rulers. The newly independent states were divided along the same boundaries as the divisions in the Spanish colonial bureaucracy.

Back in Europe, in contrast, languages and printing in the vernacular (rather than Latin, previously used for official purposes) enabled the mobilising of support for control over populations by emerging states. In Japan, the threat of conquest after 1868 triggered a process of administrative centralisation, with conscription, promotion of universal male literacy, elimination of the privileged position of the samurai, the removal of feudal controls over peasants, and subordination of local military units to a central command. Nationalism was a tool for modernisation.

Anderson identifies another type of nationalism, sponsored by governments that wanted to prevent challenges from below. This sort of "official nationalism" was important in Europe in the mid 1800s. The Austro-Hungarian empire, for example, was threatened by popular nationalism, so it sponsored its own fake nationalism. This involved rewriting of history, official propaganda and compulsory state-run education (presenting a mythical national past). Nevertheless, there was a tension in official nationalism between the myth of a single ancestor nation and the reality of an empire containing several possible nations.

The paradoxes of official nationalism were accentuated in England, where a mythical history of England was developed. It was mythical in that there was no historical English nation. For example, some of the supposedly "English" kings were from continental European dynastic houses and could not even speak English, and centuries ago residents of what is today called England had no sense of being part of a nation. Anderson notes, parenthetically:

> The barons who imposed Magna Carta on John Plantegenet did not speak "English," and had no conception of themselves as "Englishmen," but they were firmly defined as early patriots in the classrooms of the United Kingdom 700 years later.[2]

---

2 Ibid., p. 118.

There was also a tension between England as a nation and the reality of an empire. In the 1800s within the empire, aspiring colonials seeking a career in government service were blocked in their advancement. A talented, educated bureaucrat from India could never attain a position in London, nor even in the capitals of colonies in Africa such as Kenya. Anderson notes that there was a strong dose of racism in British colonial policies, but that white colonials, for example from Australia and New Zealand, faced the same blockages. The reality was an empire ruled by upper class figures at the centre, so the idea of a nation, in which all members have some sort of common membership and some level of equality, was patched on top and never fully convincing, hence the need for government sponsored efforts to foster a manufactured national myth.

After the initial development of nationalism in some parts of the world, it became a model for use elsewhere, by both insurgent movements against colonial powers in Africa and Asia and by governments to forestall chal-lenges. As a model, nationalism has been extraordinarily powerful. Anderson notes the significance of the wars between China, Vietnam and Cambodia in the late 1970s. These wars were the first between socialist states, states that were premised on international solidarity of the working classes. In practice, though, rulers found it expe-dient to encourage citizens to identify with the state rather than the working class. Anderson notes that the average Chinese peasant had no particular interest in a dispute with peoples to the south.

Anderson addresses the connection between nationalism and racism. It is commonly thought that these are
related, but Anderson notes a positive side to the emotional dimension of nationalism, namely that it is about
love for a country, not contempt for supposedly lesser
ethnicities. He points to a remarkable absence, among
writers from subjugated populations, of antagonism
towards their oppressors: they are far more likely to laud
their own culture than to denigrate others. Though there is
more to say about the connection between nationalism and
racism, it is wise not to assume they are automatically
related.

John Breuilly and nationalism as politics
John Breuilly presents a useful perspective in his book
Nationalism and the State.[3] Basically, he sees nationalism
as a form of politics, in other words as a way of exercising
power, most commonly to take control of the state. To
appreciate Breuilly's perspective, it's helpful to look first
at conventional views of nationalism that see it as associated with support for a nation, based on cultural characteristics such as language, ethnicity and customs. The
usual idea is that members of a nation may feel oppressed
by a state and seek to create a state of their own.

Breuilly says it is more the other way around. Certain
groups want to increase their power, and can do this by
challenging the state, seeking the power of a state for
themselves. They could justify their challenge by claiming

---

3 John Breuilly, *Nationalism and the State* (Manchester:
Manchester University Press, 1993, 2nd edition).

to be superior administrators or having a better set of beliefs, for example defending freedom from tyranny as in the American Revolution. However, in many circumstances it is more effective for challengers to claim to represent the aspirations of a nation. For this purpose, they then refer to an illustrious history of the nation and emphasise cultural characteristics that distinguish their group from others.

Consider Yugoslavia, a country prior to 1990 containing many different ethnic groups: Serbians, Croatians, Slovenians and so forth. After the collapse of Eastern European regimes in 1989, there was a struggle for power in Yugoslavia, eventually leading to war. Nationalism was invoked as an explanation for the breakup of the country but, looking at the process from Breuilly's perspective, actually the struggle for power was the primary driver, and national characteristics were used as a justification. This was most obvious in Bosnia, where Serbians, Croatians and Muslims (not a national group) had long lived together without difficulty. In the Bosnian war, the idea of nations seeking autonomy was the pretext for a bitter quest for power.

Breuilly takes "nationalism" to refer to "political movements seeking or exercising state power" that use a political argument with these three features: (1) there is a nation with its own special features; (2) the nation's interests and values are paramount; and (3) the nation needs to be independent.[4] The key bit of this viewpoint is

---

4 Ibid., 2.

that nationalism is all about power, in particular state power.

Another part of Breuilly's argument is that the rise of nationalism occurred along with the rise of modern states, initially in Europe and then worldwide via European colonialism. Without the state, there would be no point of nationalistic fervour. Like Anderson, Breuilly says that people centuries ago, before the rise of modern states, did not think of themselves in terms of nations. Their identification was more local.

Breuilly's analysis of nationalism is based on a wide-ranging examination of movements from around the world, including for example both unification and separation nationalism in Europe in the 1800s, anti-colonialism nationalism in India, Kenya and elsewhere, reform nationalism in China, Japan and Turkey, and nationalism after the collapse of communism in Eastern Europe and the Soviet Union.

Breuilly's perspective can be summed up this way:

Nationalism is not the expression of nationality, if by nationality is understood an independently developed ideology or group sentiment broadly diffused through the "nation." ... Rather, an effective nationalism develops where it makes political sense for an opposition to the government to claim to represent the nation against the present state.[5]

---

5 Ibid., 398.

My aim in this book is to point out the use of tactics by ruling groups to maintain their power. Breuilly's perspective meshes quite well with the study of tactics, because he's saying that the mobilisation of support for a political movement by reference to national characteristics is usefully understood as a political strategy, not as something inherent in a nation.[6]

Michael Billig and banal nationalism

In his important book *Banal Nationalism,*[7] Michael Billig gives a different perspective than Breuilly. "Banal" refers to things that are ordinary, routine and everyday. Billig argues that nationalism is not just something that is emotional, extreme and usually somewhere else, but is around us all the time even when it is unnoticed: it is banal. He gives the example of the US flag, which is hung from people's homes and printed on T-shirts. Most of these flags and flag images are treated as part of the background of daily life, yet they foster a consciousness of

---

6 Since writing *Nationalism and the State,* Breuilly's ideas have evolved. See for example "Nationalism as global history," in Daphne Halikiopoulou and Sofia Vasilopoulou (eds.), *Nationalism and Globalisation: Conflicting or Complementary?* (London: Routledge, 2011), pp. 65–83; John Breuilly (ed.), *The Oxford Handbook of the History of Nationalism* (Oxford: Oxford University Press, 2013); John Breuilly, "Nationalism," in John Baylis, Steve Smith and Patricia Owens (eds.), *The Globalization of World Politics: An Introduction to International Relations* (Oxford: Oxford University Press, 2014, sixth edition), pp. 387–400.

7 Michael Billig, *Banal Nationalism* (London: Sage, 1995).

the nation as integral to the fabric of life. Similarly, in schools around the country, children daily stand, put their hands on their hearts and together recite the pledge of allegiance: "I pledge allegiance to the Flag of the United States of America, and to the Republic for which it stands, one Nation under God, indivisible, with liberty and justice for all."

By referring to nationalism as banal, as ordinary, Billig is not saying it has no adverse consequences. As he puts it, "banal does not imply benign."[8] Banal nationalism can be toxic in its own way, blinding citizens to the assumptions underpinning the way they see the world and enabling aggression and wars.

Billig, like other writers on nationalism, notes that just a few hundred years ago very few people had any conception of themselves as members of a nation. In medieval Europe, peasants saw their world as extending only to the groups of people they interacted with and encompassing a limited geographical area without fixed boundaries. Few people living in what is today called France thought of themselves as French. In today's world, in contrast, every bit of land is assigned to one country or another and boundaries are clearly demarcated. The idea that there could be large numbers of people not attached to countries or there could be populated territory not included in a country is hard to grasp.

The contemporary way of thinking about the world is built on assumptions about membership of groups and the division of territories, assumptions that are hard to

---

8 Ibid., p. 6.

appreciate because they are unspoken and seldom articulated. When a political leader says "We must protect the French way of life," it is not necessary to spell out that "France" is being distinguished from other distinct countries and that it is reasonable to assume the existence of a "way of life" for everyone encompassed by the adjective "French" despite the vast differences in thought and behaviour between different people implicated in the term. Billig says that, "nationalism is the ideology by which the world of nations has come to seem the natural world—as if there could not possibly be a world without nations."[9]

Billig thus conceives nationalism as something more pervasive and unnoticed that the usual usages by scholars in the field who, like Breuilly, see it as mainly being manifested in challenges to existing states. Much of Billig's book is a critique of scholarship that ignores the routine and fails to examine assumptions underlying the current way the world is organised and thought about. He addresses the claims of postmodernists that national consciousness is being superseded by other forms of identity, and shows postmodernists' failure to consider banal nationalism. He provides a close critique of the work of famous philosopher Richard Rorty, showing Rorty's philosophical pragmatism is built on unacknowledged assumptions about US nationalism. Billig's many examples include several that I address in later chapters, including language and sport.[10]

---

9 Ibid., p. 37.

10 Billig's ideas have been the subject of critical attention. See for example Michael Skey, "The national in everyday life: a

What Billig calls nationalism I might call statism or countryism or country-centredness, but the terms are less important than the basic idea, namely that people think of the world as divided into countries and of themselves as members of a country or a nation.

Conclusion

There are several common themes in the books by Anderson, Breuilly and Billig. One key point is that the idea of nations is quite new, no more than two or three hundred years old. Earlier than this, and even today in many parts of the world, people have not thought of themselves as part of a nation or a nation-state. The idea that the world is divided up into geographically bounded areas, each one administered by a central government, is new historically. What seems natural today would have seemed unnatural, even incomprehensible, to earlier generations.

All three authors see the rise of the idea that people have national identities as happening in parallel with the rise of the state system. States rule over people living within territories; national identity helps make this seem natural and inevitable rather than arbitrary and open to challenge.

Another key point is that effort is required to get people to think in terms of nations, states, borders, citizen-

---

critical engagement with Michael Billig's thesis of *Banal Nationalism*," *Sociological Review,* Vol. 57, No. 2, 2009, pp. 331–346, and Michael Billig, "Reflecting on a critical engagement with banal nationalism—reply to Skey," pp. 347–352.

ship and all the other facets of the system of states. Sometimes the efforts are strenuous and obvious, such as during wartime, but more commonly the usual ways of thinking about the world are reinforced by education, media and everyday rituals.

Finally, it is important to recognise that the state system is a power system. It is political, in the sense of involving the exercise of power. Many individuals and groups have a stake in the way the world is organised and resist those who promote alternatives. One of the key uses of power is to encourage people to think that the system is natural and that alternatives are impractical.

The body of writing about nationalism and states is enormous and there is no possibility of even trying to summarise it. My goal in *Ruling Tactics* is to point to ways in which governments and their supporters encourage people to think in terms of countries and from the point of view of governments. In doing this, I am drawing on several sources. One is the body of research about nationalism, and Billig's *Banal Nationalism* is as close as any treatment to my starting point. Another source is the analysis of strategy and tactics in the social world; James Jasper's book *Getting Your Way* is the pioneering treatment, showing how social dynamics can be analysed in terms of strategy.[11] Finally, I have drawn on my own study of tactics against injustice, which offers a framework for understanding the methods used by powerful groups to reduce outrage over injustice, and which can be used more

11 James M. Jasper, *Getting Your Way: Strategic Dilemmas in the Real World* (Chicago: University of Chicago Press, 2006).

generally to look at tactics adopted by rulers.[12] My aim is to use a range of topics to illustrate how, by looking at familiar things in different ways, it is possible to recognise tactics that help maintain systems of rule and to imagine ways to take action towards alternatives.

---

12 Brian Martin, *Justice Ignited: The Dynamics of Backfire* (Lanham, MD: Rowman & Littlefield, 2007). See, more generally, "Backfire materials," http://www.bmartin.cc/pubs/backfire.html.

4
Crime

Murder, theft, assault, burglary—these are staples of news coverage. People hear a lot about crime, and nearly everyone thinks it's a bad thing. Yet there are huge differences in ways different actions are labelled as crime and in the attention they receive.

The first distortion is that most attention is given to low-level crime, the sort that hurts a few people and is carried out by relatively powerless individuals. This includes many murders, which attract a lot of attention. Indeed, so potent is murder for attracting attention that it has become a staple of news coverage as well as crime novels and television shows: think of Agatha Christie and CSI and many others like them.

Murder is usually thought of as something done by an evil person, who needs to be tracked down, proven guilty and punished. Most despicable of all is the serial killer who preys on victims over a period of years.

Yet there is another sort of crime that usually escapes the spotlight, and those responsible are seldom identified or exposed, much less ever prosecuted and convicted. This is crime by those with a lot of power.

Let's start with corporate crime. Corporate executives may enact policies that predictably kill people, sometimes large numbers of people. They may hide evidence showing how many people are dying due to their actions.

A classic example involved the Ford Pinto. As exposed in a classic 1977 article in the magazine *Mother Jones,*[1] Ford engineers and executives knew about a fault in the fuel system: collisions to the rear end of the vehicle could easily rupture the fuel system, leading to fire and potentially to death of the occupants. Ford already had a patent for a safer gas tank, but to save money—Ford was then in competition with Volkswagen for the US small-car market—the company retained the dangerous tank, and for years lobbied against government standards that would have mandated a safer tank. Hundreds of people died from burns, and Ford settled numerous damage claims out of court. The company's internal cost-benefit analysis showed that paying damage claims was cheaper than putting in the safer tank. Was this a crime? Technically not, because auto manufacturers had lobbied against any provision in the Motor Vehicle Safety Act providing for criminal sanctions for selling unsafe cars. However, it is not something that any company would want to admit, much less advertise.

On a vastly greater scale are the actions of tobacco companies. Executives know that smoking cigarettes leads to the illness and premature death of a great number of smokers. Furthermore, the companies carried out research of their own that showed the dangers while denying them publicly. They fought regulations tooth and nail.[2]

---

1 Mark Dowie, "Pinto madness," *Mother Jones,* September/ October 1977.

2 Stanton A. Glantz, John Slade, Lisa A. Bero, Peter Hanauer and Deborah E. Barnes, *The Cigarette Papers* (Berkeley, CA: Univer-

The movement against smoking has been remarkably successful in exposing the actions of tobacco companies. Fines of hundreds of billions of dollars have been imposed. Yet the companies still do all they can to expand sales around the world and to resist regulations, such as plain-paper packaging, that discourage smoking.

If ever there was an industry causing mass death, it is the tobacco industry. One estimate is that a billion people may die this century due to tobacco-related diseases. How many tobacco company executives have gone to jail for their responsibility?

Another example is the scandal involving the Australian Wheat Board (AWB), a government agency (privatised in 1999) with a monopoly on selling Australian wheat. Between 1991 and 2003, Iraq was subject to UN sanctions that blocked the import of many items. The AWB was eager to make sales to Iraq during this time—so eager that extra payments were paid to dealers, money that went to the regime in violation of the sanctions, right up until the time the Australian government sent troops as part of the 2003 invasion of Iraq. A$290 million in bribes was involved, a huge support for Saddam Hussein's regime. The story eventually broke in Australia, and there was an inquiry and recommendations for criminal charges, but the police did not proceed: no AWB officials were prosecuted for crimes, much less went to jail.[3]

---

sity of California Press, 1996); Robert N. Proctor, *Golden Holocaust: Origins of the Cigarette Catastrophe and the Case for Abolition* (Berkeley, CA: University of California Press, 2012).

3 Caroline Overington, *Kickback: Inside the Australian Wheat*

Despite payments by the AWB and other importers, the sanctions against Iraq were remarkably effective, not in hobbling Saddam Hussein's grasp on power, but in harming the Iraqi people. Due to shortages of sanitation equipment, medicines, and other vital materials, the death rate due to malnutrition and disease soared. Perhaps one or two million Iraqis died as a result of the sanctions. In a famous quote, US secretary of state Madeleine Albright was asked whether the sanctions could be justified given the death of half a million Iraqi children. She answered, "I think this is a very hard choice, but the price—we think the price is worth it."

Some commentators have judged the sanctions against Iraq to constitute genocide: actions taken knowingly leading to mass death in a target population.[4] No one was ever charged with a crime.

The 2003 invasion of Iraq, led by the US government, was not approved by the UN Security Council. In the eyes of many legal scholars, it was an illegal war, yet no one responsible was ever charged.

Board Scandal (Sydney: Allen and Unwin, 2007).

4 Geoff Simons, *The Scourging of Iraq: Sanctions, Law and Natural Justice,* 2nd ed. (Basingstoke: Macmillan, 1998). For a discussion of the shortcomings of international governance in this case, see Joy Gordon, *Invisible War: The United States and the Iraq Sanctions* (Cambridge, MA: Harvard University Press, 2010), pp. 221–230.

Journalist James Risen has told of corruption in the aftermath of the invasion.[5] To prop up the collapsed Iraqi economy, masses of US cash were flown from the New York Federal Reserve Bank to Iraq. The amounts were so great, even in US $100 bills, that entire cargo planes were filled with the cash, ultimately $12 to $14 billion. To be handling so much cash was a temptation for everyone involved, including US soldiers who were supposed to count or distribute some of the money. Accounting procedures were so lax that billions of dollars went missing, no one knows where—at least no one in official places. Information was pieced together indicating that a couple of billion dollars were stored in Lebanon on behalf of corrupt Iraqi government figures. Although provided with addresses, US officials showed little interest in pursuing the cash or the criminals. Apparently it was all too embarrassing for US figures involved in the operation.

Crime and the law
Breaking the law is an offence, and not breaking the law is okay, right? Well, it depends. Some laws are broken so often and enforced so infrequently that few are concerned. Laws against jaywalking are an example, in places where pedestrians routinely cross the street anywhere they please. So is photocopying or scanning a book that's in copyright. Cash-in-hand payments to tradespeople enable tax avoidance. Technically, in many places, these actions are illegal, but no one bothers about them.

---

5 James Risen, *Pay Any Price: Greed, Power, and Endless War* (Boston: Mariner Books, 2015).

Then there are legal loopholes, which are ways to cheat legally. In the US tax code, legislators have written in hundreds of special exemptions that apply to a single business or individual.[6] Corporate lawyers search for loopholes to minimise the tax their companies pay. Tax havens—countries imposing little or no company tax—are legal, and expressly designed to help multinational companies avoid tax in the countries where they do most of their business.[7]

There's an old saying that the golden rule means "He who has the gold makes the rules." In other words, those with wealth have influence over how the law is written and enforced. Consider an example: a company owner decides to fire all the employees and hire new ones at lower wages. In some places, this is legal; in other places, it's not legal, but government regulators would not bother to prosecute. In such circumstances, the main restraint on this sort of action is the organised action of workers and their supporters, for example via a work-in or a blockade.

So there are two ways to think about crime and the law. One is the technical one: something is only a crime if it's against the law. The other is the social one: something is a crime if it defies widespread community expectations for fair and ethical behaviour.

---

6 Donald L. Barlett and James B. Steele, *America: Who Really Pays the Taxes?* (New York: Simon & Schuster, 1994). They also describe a multitude of ways the US tax system has been manipulated to serve the rich.

7 See chapter 11, "Trade deals and tax havens."

If someone is homeless and sleeps on a park bench, is this a crime? If someone passes out leaflets in a shopping centre, is this a crime? Technically, these behaviours may or may not be legal, depending on local laws. Socially, observers will differ in their views about homeless people or leafletting: whether something is a crime depends on the way you think about the behaviour and about the law.

State crime
State crime refers to crimes committed by governments and government agents.[8] However, in many cases, actions by governments are treated as above the law. An example is torture. Nominally, in nearly every country in the world torture is considered a crime, but seldom is anyone charged or convicted of committing torture, least of all by the governments that sanction it.

Consider first the manufacture of equipment used for torture, everything from thumbscrews to electroshock batons. This is a huge industry.[9] There are "security fairs" held in countries around the world displaying the latest

---

8 Jeffrey Ian Ross, ed., *Controlling State Crime*, 2nd ed. (New Brunswick, NJ: Transaction Publishers, 2000); Jeffrey Ian Ross, ed., *Varieties of State Crime and Its Control* (Monsey, NY: Criminal Justice Press, 2000); Dawn L. Rothe, *State Criminality: The Crime of All Crimes* (Lanham, MD: Rowman & Littlefield, 2009); Dawn L. Rothe and Christopher W. Mullins (eds.), *State Crime: Current Perspectives* (New Brunswick, NJ: Rutgers University Press, 2011). See also the discussion of state terrorism—a type of state crime—in chapter 7.

9 See publications of the Omega Research Foundation, https://omegaresearchfoundation.org/publications/.

equipment for surveillance and control. There is also a well-developed system for training personnel in "advanced interrogation techniques," a euphemism for torture. Yet it is rare, indeed almost unheard of, for anyone involved in what should be called the torture trade to be considered a criminal.

Then there is torture in practice. Governments know it is going on, but usually will do nothing unless there is adverse publicity, and naturally enough they usually avoid publicity if at all possible.

In the aftermath of the 2001 invasion of Afghanistan and the 2003 invasion of Iraq, there were reports about torture in US facilities. There wasn't much concern until photos from Abu Ghraib prison in Iraq became public in 2004. These showed Iraqi prisoners being piled naked on top of each other, a hooded Iraqi prisoner in a stress position apparently in fear of being electrocuted, and an Iraqi prisoner being threatened by a dog, among other gruesome images. It was only because of the massive publicity generated by these photos that a few US prison guards were charged with crimes. However, the US government avoided the word "torture," referring instead to "abuse," and the US mass media followed suit. The government implied actions by guards at Abu Ghraib were their own initiative, ignoring evidence of higher responsibility.[10]

---

10 Jennifer K. Harbury, *Truth, Torture, and the American Way: The History and Consequences of U.S. Involvement in Torture* (Boston: Beacon Press, 2005); Alfred W. McCoy, *A Question of*

Abu Ghraib prison torture was an anomaly, not because it involved torture, but because it was exposed. It was business as usual in the sense that higher officials escaped any censure.

Then there are more routine forms of torture. In US prisons, it is commonplace for prisoners to be subject to treatment that fits usual definitions of torture. Supermax prisons, in which prisoners are kept in isolation most of the time, serve as a form of torture, using the techniques of sensory deprivation pioneered by the British in Northern Ireland.[11] Restraint chairs and electroshock weapons are regularly used to control resistant prisoners, and guards may knowingly allow prisoners to assault each other.[12] It would be possible to argue that there are more crimes committed against prisoners in US prisons than the prisoners ever committed on the outside, especially considering that many are in prison for victimless law-breaking such as using drugs. Yet the guards responsible for direct assaults on prisoners are almost never charged with crimes. Even less likely is it that politicians and planners who design prison systems will ever be thought of as criminals.

Torture: CIA Interrogation, from the Cold War to the War on Terror (New York: Metropolitan, 2006).

11 Carol Ackroyd, Karen Margolis, Jonathan Rosenhead and Tim Shallice, *The Technology of Political Control* (Harmondsworth: Penguin, 1977).

12 On one aspect of this, see Joanne Mariner, *No Escape: Male Rape in U.S. Prisons* (New York: Human Rights Watch, 2001), https://www.hrw.org/legacy/reports/2001/prison/report.html.

To summarise several points covered so far, most attention in the media is to crimes of individuals and to crime by people lower down the social hierarchy. Corporate crime is neglected because it is systemic and those most responsible are top executives. In practice many laws are broken all the time with impunity, and powerful and influential groups are able to influence lawmakers and prosecutors so that their shady operations, such as tax avoidance, are technically legal. One special category is state crime, which is crime by governments and their agents. It receives little public attention and is seldom punished.

Collins on crime

Sociologist Randall Collins provides a valuable insight into the dynamics of crime.[13] He notes that conservatives explain crime as an individual failing, due to genetics or poor character: their solution is punishment. However, this approach doesn't solve the problem and is best understood as a moral and political position.

Liberal explanations focus on crime cultures, including poverty, with the solution being rehabilitation. These explanations are not satisfactory either, because many poor people are not criminals and many rich ones are.

Radical explanations see crime as a category of behaviour that is labelled as criminal, with convictions produced by the law-enforcement machinery. From this viewpoint, laws create crime, especially victimless

---

13 Randall Collins, "The normalcy of crime," in *Sociological Insight: An Introduction to Nonobvious Sociology* (New York: Oxford University Press, 1982), pp. 86–118.

lawbreaking such as illicit drug use, thereby fostering the creation of criminal cultures. This explanation doesn't work well for property and personal crimes such as theft and assault.

The class-conflict model, derived from Marxism, sees crimes as due to class relations, especially the existence of private property. However, socialist societies still have crime; indeed, they create new categories of it, especially crimes against the state.

Collins notes that in Denmark in 1944, there were no police. Property crimes greatly increased but crimes against people stayed about the same.

Collins' own preference is a picture derived from Emil Durkheim, one of the founders of sociology: crime and punishment serve as a bond for the rest of the community. A stratified society, in which some groups have far more wealth and power than others, can be unified by rituals, and one potent ritual is punishment of those labelled criminals. This helps explain the attraction of murder mysteries. Collins says that in power struggles, there are plenty of actions that can provide offence. Some of these are criminalised—turned into crimes by laws and expectations for punishment—and thus provide opportunities for ceremonies of punishment that dramatise the moral feelings of the community. Each type of society has its own forms of crime.

Tactics: dilemmas for the state
For government leaders and supporters, the topic of crime contains opportunities and dangers. Fears about crime can

be drummed up, but there is a risk the spotlight might be turned on crimes by those with power and wealth.

The first tactic used by governments to foster a preferred orientation towards crime is exposure. Government leaders direct attention towards low-level crime, and crime by individuals, and the media usually are willing accomplices. Murder—usually involving killing of one individual by another—has become a topic that, to many people, is fascinating. The mass media report on murders, some of which become ongoing sagas. The case of O J Simpson, a famous US gridiron star accused of murdering his wife and a friend, attracted enormous media attention. Fictional treatments of murder, in novels and television shows, are also popular. It's as if news media and the entertainment industry are saying, "Look, here's what you should be concerned about."

The attention to individual crime—murder, yes, and assault, robbery and embezzlement—serves to create a perception that crime is due to bad people. There is correspondingly little attention to state and corporate crime, including the arms trade, illegal wars, and sales of dangerous products such as pharmaceutical drugs. The crimes by states and corporations cause far more deaths than individual murders but in comparison receive little attention.

A similar disparity occurs with the second tactic, valuing. This doesn't mean valuing crime, of course, but rather valuing efforts against crime. The police, courts and various agencies are commonly portrayed in news stories and entertainment as the good guys, taking up the noble cause of cracking down on drug dealers, robbers, hooligans and welfare cheats. Valuing comes into play in the resources given to enforcement agencies. For example, huge amounts of money are provided to anti-terrorism bodies but comparatively little to agencies targeting high-level white-collar crime.

The third tactic is to explain the government's efforts against crime, and why they are the right ones and effective. This might involve statistics on crime rates, arrest rates, expenditures on policing, and so forth. These accounts of anti-crime efforts normally ignore questions of what should count as crime and whether the most damaging types of crime are being addressed. The figures include, typically, murder, assault, burglary and so forth, and omit a separate classification for state crime. Expla-nations of crime-fighting are sometimes designed to placate the public by indicating that everything is under control but sometimes designed to stimulate support for greater expenditures. This can be a delicate balancing act. Alarms about escalating criminal activity can scare the public and provide support for greater spending on prisons and policing, but these at the same time send a signal that the government is not doing its job of protecting the population. In either case, the most important message is what is assumed, namely that crimes by individuals, especially those with less power, are of primary concern and that institutionalised state and corporate crime is off the agenda.

To provide credibility to the government's policies, it is useful to have endorsements, which constitute the fourth tactic. Endorsements can come from police, politicians, government officials, media, experts or celebrities, among others. The basic line is normally is that the government is doing the right thing, which might be keeping crime under control or expanding its efforts against a new type of crime, due for example to the drug ice or cybercriminals.

It is a different story when it comes to experts who present a non-standard view about crime, a story pointing to fundamental flaws in crime control. This will differ from country to country, but consider the idea of restorative justice. In countries like the US, convicted criminals are incarcerated and seldom provided extensive support for rehabilitation: the dominant approach is retribution, namely punishment. It is so standard that it is simply taken for granted in most commentary by police, prison officials and politicians. There is another approach, called restorative justice, involving meetings between offenders and those harmed and finding a mutually agreed response, often with apology and restitution (payment, community service and sometimes incarceration). The motivating philosophy behind restorative justice is to reintegrate offenders into the community, not to brand them for life.[14]

There are many ardent advocates for restorative practices but, in the US at least, they receive little public attention. Endorsements of the standard retribution model are given far more visibility.

Consider also the radical critique of US prison policy, for example by eminent Norwegian criminologist Nils Christie, author of *Crime Control as Industry: Towards*

---

14 John Braithwaite, *Restorative Justice and Responsive Regulation* (Oxford: Oxford University Press, 2002).

Gulags, Western Style.[15] There are other critics too, within the US, who advocate abolishing prisons. These alternatives receive little visibility in the media.

The criminal justice system—what critics might call the criminal injustice system—offers ample rewards for those who emphasise the usual sorts of low-level crime. There are many jobs in the system, in what has been called the prison-industrial complex, for building prisons, running police operations and a host of related activities.[16] In comparison, there are relatively few rewards for those pushing for alternatives such as restorative justice and prison abolition.

Conclusion
If Randall Collins' ideas about the role of crime in society are accepted, then it is predictable that in just about any society some actions will be labelled criminal, stigmatised and penalised. However, this can be done in various ways, with differing social and political effects. In the restorative justice approach, damaging behaviours are dealt with by community-based efforts to reintegrate the offender into a meaningful social group.

For rulers, though, there are two great temptations. The first is to use the advantages of power to commit crimes or, even better, to set up the rules so that personal wealth can legally be extracted from the population. The

---

15 Nils Christie, *Crime Control as Industry: Towards Gulags, Western Style* (London: Routledge, 1994).

16 Joel Dyer, *The Perpetual Prisoner Machine—How America Profits from Crime* (Boulder, CO: Westview Press, 2000).

second temptation is to raise the alarm about other sorts of crime, by enemies or by people lower in the social hierarchy. A parallel double process of persuasion is involved: hide high-level crime or make it legal, acceptable and even laudable, and at the same time encourage people to project their fears and anger about crime onto scapegoats.

The result of these temptations and tactics is obvious in media coverage (encouraged by government priorities, for example funding police and anti-fraud agencies) and hence in everyday conversations. If the size of a crime, or the proceeds of rules that enable unfair distribution of wealth, were the determinant of attention, hardly anyone would get excited about low-level theft when corporations and governments are extracting, legally or not, billions of dollars from the populace. When it comes to crimes of violence, if the scale of devastation and death were the determinant of attention, then media coverage would concentrate on state terrorism, not the small-scale efforts of non-state groups.

To challenge the dominant narrative about crime is difficult, but can be done. It involves continual exposure of the techniques used by governments and others to direct attention to individual criminals, and efforts to promote alternatives. There are many people doing this, in all sorts of ways. Useful lessons can be learned from efforts to challenge the so-called war on drugs: exposing its sordid origins and harmful effects, showing the rationality and publicising the beneficial effects of decriminalising drugs (as in Portugal), and fostering sensible ways to reduce the harmful effects of addiction (rather than assuming legalisation on its own is sufficient).[17] Proponents of harm reduction want to treat drug use as a social and health problem rather than a legal and policing problem. Similar efforts can be taken on other issues in which the "crime problem" is used to serve vested interests.

17 Johann Hari, *Chasing the Scream: The First and Last Days of the War on Drugs* (New York: Bloomsbury, 2015).

5
Sport

It's the year of the Olympic Games. For many fans, this is one of the highlights of the sporting calendar. Even those who do not follow sports may tune into the spectacular opening ceremony.

The modern Olympics supposedly were set up for noble purposes: instead of waging war, nations could engage in peaceful, healthy competition. From the start, though, the games were driven by baser considerations, including nationalism and, later, commercialism.[1]

---

1 Useful treatments include Robert K. Barney, Stephen R. Wenn and Scott G. Martyn, *Selling the Five Rings: The International Olympic Committee and the Rise of Olympic Commercialism* (Salt Lake City, UT: University of Utah Press, 2002); Jules Boykoff, *Celebration Capitalism and the Olympic Games* (New York: Routledge, 2013); Jules Boykoff, *Power Games: A Political History of the Olympics* (London: Verso, 2016); Richard Espy, *The Politics of the Olympic Games* (Berkeley: University of California Press, 1979); Christopher R. Hill, *Olympic Politics* (Manchester: Manchester University Press, 1992); John Hoberman, *The Olympic Crisis: Sport, Politics and the Moral Order* (New Rochelle, NY: Aristide D. Caratzas, 1986); Jeffrey Segrave and Donald Chu (eds.), *Olympism* (Champaign, IL: Human Kinetics, 1981); Alan Tomlinson and Garry Whannel (eds.), *Five-Ring Circus: Money, Power and Politics at the Olympic Games* (London: Pluto Press, 1984).

The nationalist bias is apparent in several features. Athletes compete as representatives of their country of citizenship. In individual events, no more than three competitors are allowed from any country. In team sports, such as basketball, each country can have only one team. So the Olympics, rather than being a genuine world championship of the best athletes, are constrained by the artificial barriers of citizenship. Team sports in particular can become surrogates for international rivalries.

In the opening ceremony, watched by billions around the globe, athletes march around the host stadium in national teams. It is a special honour for an athlete to lead the team, carrying the country's flag.

In most media coverage of the Olympics, a country's media concentrate on the progress of their "own" athletes, namely the ones representing their country. Viewers are encouraged to identify with these particular athletes. For example, in the 2000 Sydney Olympics, Cathy Freeman, a sprinter in the 400 meters and a prominent Indigenous Australian, was favoured to win. She was chosen that year to light the Olympic flame. When she won the final in her signature event, there was rejoicing throughout the country, with saturation media coverage both encouraging and responding to this popular interest. Many viewers saw Freeman's victory as not just a personal achievement but as representing Indigenous success and, more generally, an achievement for the whole country, especially given these games were in Australia.

Hosting the Olympics is treated as a matter of national prestige, as well as kudos for the city. Government officials use all sorts of persuasive means, including bribery, to encourage Olympic committee members to support their bids to host the Olympics. Politicians and entrepreneurs in cities and countries where the Olympics are held use the opportunity to sell their preferred image, perhaps as a responsible member of the international community (for example, Beijing) or a desirable tourist destination (for example, Sydney). A winning bid to host the games is widely seen as a diplomatic triumph, despite the huge costs and headaches in getting the venues ready in time.

Behind the scenes at Olympic host cities, governments carry out various forms of civil and social engineering in order to present a positive picture to the world. This includes moving homeless people out of city centres, bulldozing homes, suppressing dissent and creating huge facades.

Meanwhile, among the athletes, every attempt is made to foster a clean image. Competitors, in their quest for Olympic gold, will make all sorts of sacrifices, and for some this includes performing while injured, using banned drugs and using unethical techniques to undermine opponents.

The Olympics are officially presented as a moral paragon, as a unifying enterprise for the world. In practice, Olympics politics represent one of the worst models of compromise and lack of principle. In order to enable participation, nearly every government, no matter how dictatorial and corrupt, is allowed to send a team. Thus oppressive regimes can bask in the reflected glory of having their chosen athletes compete. For some governments, participation is restricted to those considered acceptable. For decades, numerous governments allowed only men to compete, and persecuted minorities are commonly excluded. Olympic officials seldom intervene in decisions made by national committees. In these ways, the Olympic movement panders to nationalism.

The Olympic Games have often been used as tools in international diplomacy. In 1980, many governments boycotted the games in Moscow as a protest against the Soviet invasion of Afghanistan. In some places, Australia among them, athletes were placed in a quandary. Should they follow the recommendation of their government and boycott the games, or instead attend anyway in order to achieve what for many is a once-in-a-lifetime opportunity to participate in the world's most prestigious sporting competition? In 1984, Soviet-bloc governments held a payback boycott of the games in Los Angeles.

The Olympics have also become highly commercial, especially with the rise of television coverage in the 1950s and 1960s, providing huge revenues to the International Olympic Committee and leading to transformation of the games into an ever greater spectacle.

Originally, Olympic athletes were required to be amateurs. This gave an advantage to members of upper classes who had access to facilities and leisure time for training. As the Olympics became more prestigious, some governments and athletic organisations gave support to their athletes in ways that got around the amateur rule. In the Soviet bloc, elite athletes were de facto professionals with sham jobs. In the US, athletic scholarships at universities, combined with soft study options, enabled many athletes to train almost like full-time professionals.

Furthermore, many received covert payments or benefits, so this era was sometimes called "shamateurism." The ending of the amateur requirement meant these forms of hypocrisy were avoided, though at the expense of the original Olympic ideal.

In the Soviet bloc, scientists were involved in designing training for national teams. The East German team was highly successful, producing many world champions, and was also notorious for the widespread use of banned drugs. The quest for Olympic gold was so strong that athletes in many other countries also used drugs.

Drugs are only one way to seek a competitive advantage. In several countries, national training centres undertake research to support elite athletic performance. In the 1976 Montreal Olympics, the Australian team did not obtain a single gold medal, a dismal performance in the eyes of political leaders who saw athletic success as a source of national pride. In response, the government set up the Australian Institute of Sport to undertake research and oversee training of elite athletes. This was modelled on the Eastern European efforts, but without the emphasis on drugs. The AIS has studied training regimens, psychology, special technological aids in training, coaching techniques and other areas. It has been one factor in the later successes of the Australian team, putting it ahead of larger countries on a per-capita basis.

The rhetoric of sport sometimes proclaims that the aim is participation, not winning, but in practice the emphasis is on victory, as in US football coach Vince Lombardi's famous saying "Winning isn't everything, it's the only thing." The emphasis on winners, and on elite athletes, is obvious in media coverage. Olympic coverage is on the finals and on medallists, especially gold medallists. In many countries, coverage is selective, with attention given to athletes from the country in question. A local hero might be followed through the heats of an event, but if beaten, coverage switched to other events. Viewers who want to watch the "best in the world" may be frustrated by coverage oriented to national competitors.

Then there are the unofficial medal totals, listing the number of gold, silver and bronze medals obtained by athletes from different countries. In 2012, the countries with the most total medals were the US, China and Russia. However, further down the list, many people in countries such as Algeria and Guatemala were proud that a single competitor from their country received a medal.

Media coverage and medal totals encourage identification with a national team, and with a country. Flags are in abundance, and national anthems played for winners. These processes encourage citizens to identify with the elite athletes from their own country. (Immigrants often have conflicted loyalties.)

When citizens identify with Olympic athletes from their countries, many important differences are obscured. Just because runners or shooters on an Olympic team do well does not mean ordinary citizens from the country are any good at running or shooting. They might be, but many of them might be unable to run more than a short distance and never have used a rifle. Success in the Olympics can make viewers feel like winners, by proxy.

Olympic athletes must be highly dedicated to achieve world-class performance levels. This typically means spending hours per day for years on end. This sort of commitment is uncommon. Viewers can bask in the illusion that dedication by athletes has some spin-off association with dedication by themselves or others in their societies. This might be true, but often isn't.

The Olympics, like most sports in other contexts, are presented as clean, honest, fair competitions, in which those with superior capabilities are victorious. Sports are widely seen as separate from the messy and corrupt practices found elsewhere in society—they provide an escape into an ideal world. This illusion is sometimes threatened by the behaviour of athletes, for example when they are discovered cheating or when committing crimes in their life outside the sporting arena. These violations of the image of sport as pure are seen as especially objectionable, and constitute one reason why the crusade against drugs in sport is unrelenting: sport must be seen to be fair so the illusion of a pure, separate world can be maintained. Governments like to be associated with the image of sporting success—as long as it's a clean image.

Other sports

National identification is promoted via sports such as cricket and rugby in which there are national teams, so it sometimes can seem like countries are competing against each other when actually only teams representing countries are competing. Commentators say "India defeated England" rather than "The cricket team representing India defeated the cricket team representing England." Many fans identify with national teams.

In individual sports, like golf and tennis, there is some identification based on country of origin. That tennis star Rafael Nadal is from Spain can be a source of pride for Spaniards, but this is minor compared to what happens with the World Cup.[2] Football—called soccer in the US— seems to arouse tremendous passions, and nationalism is an important component of this emotional process. Football is indeed the "world game"—the US baseball finals are misleadingly called the world series—so every national team carries the hopes of many of its citizens.

The World Cup is broadcast internationally, and is the ultimate football competition. Normally, fans will back a local team, but when it's time for the World Cup, these parochial attachments are set aside in a bigger type of parochial partisanship, identification with the national team. Many athletes see their greatest achievement as playing in the World Cup, especially in the finals.

It should be noted that women's football is insignificant in audience ratings compared to the men's game. Many competitive sports remain male dominated in terms of prominence. Patriarchy influences sport in various ways, intersecting with nationalism, commercialism and other factors.[3]

---

2 David Goldblatt, *The Ball is Round: A Global History of Football* (London: Viking, 2006).

3 On the politics of football, see Gabriel Kuhn, *Soccer vs. the State: Tackling Football and Radical Politics* (Oakland, CA: PM Press, 2011).

The arbitrariness of sporting attachments

Most sports fans develop strong commitments to particular players or teams. When fans support a local player or team, loyalty is usually based in a sense of community, in which the team is treated as a representative of the locality, city, region or country.

Very few fans can observe games dispassionately, not caring who wins but simply observing a game of skill. Instead, the games involving favoured players or teams receive far more attention. If a sport is not played locally and is not widely established internationally, few fans will have any interest in it. For example, Australian rules football has a limited following in China, India or Russia.

It may seem logical that fans will support the local team, especially when the players are local identities, perhaps even meeting with the fans. However, most fan identification with players is vicarious, through watching their team, not by personally interacting with them.

The arbitrariness of these loyalties is shown when players are brought in from other parts of the country or the world. A US basketball player who joins an Australian team usually has no prior connection with Australia, yet is eagerly adopted by local fans as part of *their* team. Players and coaches are traded and transferred, basically as commodities, but only occasionally does this alienate fans. It seems that the name of the team is enough to inspire loyalty to it.

This is apparent in baseball in the US, where loyalty is most commonly to the team with a city's name, for example the Chicago Cubs or the New York Yankees. When a team moves to another city, as when the Dodgers moved from Brooklyn to Los Angeles, city-based loyalties usually trump loyalties to the players. In any case, players are regularly traded, so teams are not composed of local athletes but simply of players who have been made part of a team with a local name.

The same applies to international sport. Players are often born and bred in the country they represent, but this not essential. There are plenty of cases in which players change their citizenship in order to join a team in their adopted country. Such players are usually welcomed with open arms as one of *our* athletes. Assignment of loyalties is more about the label than about any deep connection to the country or its institutions.

Alternatives

International sporting competitions, such as the Olympics and the World Cup, seem so natural that it can be hard to imagine any other way of doing things. Therefore it is worth noting some possible alternatives, not because they are likely or even desirable, but to highlight assumptions about sport.

One alternative is simply to abolish all elite international sporting competitions. Instead, emphasis could be placed on mass participation in health-promoting and socially engaging sport and physical activity. Research shows that physical activity is a reliable way of improving happiness—more reliable than watching sporting competitions on television, for example—and there are health benefits too.[4] So if the aim is to improve gross national happiness, rather than gross national product, then widespread participation in sport is an obvious candidate.

Another option is to set up sporting competitions on a different basis, so national identification is limited. In the 1920s and 1930s, there were a number of "workers' games" run as alternatives to the Olympics. In these games, competitors did not represent countries. The orientation was to achievements by members of the working class, at a time when many elite athletes were from privileged backgrounds.[5]

Yet another option is cooperative games.[6] An example is football with an added rule: when a player scores a goal, this player joins the opposing team. When players switch sides during a game, winning becomes a side issue, because it is not even clear exactly who has won. This sort of rule undercuts the competitive dynamic and orients players to enjoying the game rather than

---

4 John J. Ratey with Eric Hagerman, *Spark: The Revolutionary New Science of Exercise and the Brain* (New York: Little, Brown, 2008).

5 Boykoff, *Power Games,* pp. 60–65; James Riordan, "The Workers' Olympics," in Alan Tomlinson and Garry Whannel (eds.), *Five-ring Circus: Money, Power and Politics at the Olympic Games* (London: Pluto, 1984), pp. 98–112.

6 Terry Orlick, *Cooperative Games and Sports: Joyful Activities for Everyone,* 2nd edition (Champaign, IL: Human Kinetics Press, 2006). More generally on the advantages of cooperation, see Alfie Kohn, *No Contest: The Case against Competition* (Boston: Houghton Mifflin, 1986).

winning. Many different sorts of cooperative games have been devised and played. Few of them would serve as vehicles for accentuating nationalism. Indeed, cooperative games might actually help break down national identification, as players from different countries helped each other in joint endeavours.

Tactics promoting nationalism in sports

There are several routine methods that promote nationalistic thinking and fervour in sport.[7] First is *exposure*: the sports, and athletes identified with countries, need to be publicised. Commonly this happens via the media, for example the worldwide coverage of the Olympics and the World Cup. Note that only some sports are publicised to a great extent. Many sports and athletes languish in obscurity, or have very limited followings. It is interesting that some of the most widely publicised competitions, with global coverage, involve athletes representing countries.

Second is *valuing*: the sports and athletes need to be seen positively. This is almost always the case for sport. Only a few sports, such as boxing, are stigmatised in some circles. Elite athletes as a group are highly esteemed, though some individuals fall from grace, especially those exposed as cheats or who commit serious crimes. By and large, athletes are seen as dedicated and talented, and are lauded for their achievements.

In 2014, 26-year-old Australian cricketer Philip Hughes was killed when hit in the head by the cricket ball while batting. This led to a huge outpouring of grief, aided

---

7 See chapter 1 for the framework used for this exposition.

by saturation media coverage. For example, some major newspapers devoted six or more pages to the story for day after day. It was the biggest such public grieving spectacle since Princess Diana died in 1997. Hughes had played on the national team and was well known to anyone who followed Australian cricket, though he was not the country's most prominent cricketer. This episode showed a confluence of valuing processes: it involved a sport that many in Australia have seen as the traditional national sport, and one in which the Australian team has often been the world's best, a young player seen as exemplary in dedication to his craft, and a sudden drama and tragedy, ideal for media coverage. It should be noted that some letter-writers were sceptical of giving so much adulation to a sportsperson, and pointed out that other people, who had made greater sacrifices to serve the community, had died without much media coverage. Perhaps a key factor was that Hughes contributed towards a sense of national identity, at least for those who followed cricket.

Nationalistic thinking is promoted using various arguments that give a *positive interpretation* of country-identified sporting competition. There is the trickle-down argument that elite sporting success will be an inspiration for others in the country, the economic argument that tourism and trade will benefit from international recognition, and the status-related argument that international prestige is tied to involvement in and success in international sport.

Another key tactic is *endorsement* of international sport by governments and national sporting bodies. This works in two ways: governments and sporting bodies endorse participation in international sport, and by competing with others implicitly endorse other teams and governments. This routine endorsement is usually unnoticed, only coming to attention when challenges are made.

From the 1960s through the 1980s, as South Africa's apartheid government faced increasing opposition to its racist policies, it sought international validation through its sporting teams. Opponents of apartheid protested against events involving South African teams. For example, in the early 1970s, there were protests in Australia and New Zealand against matches with the visiting the South African rugby team. The point here is that national teams serve as de facto ambassadors of governments, and as emblems of national pride: endorsement of the team is assumed.

The final tactic is *rewards* for joining in the glorification of athletes representing countries. The athletes themselves receive several types of rewards: the satisfaction of achievement at the highest level (being good enough to be selected for a national team is impressive), the prestige of being a winner at the international level, and occasionally financial returns from endorsements and career opportunities.

Companies can gain by associating themselves with sports. A few are involved with sports equipment, such as running shoes; others attach themselves to teams or prominent athletes through sponsorship deals; yet others benefit when a country hosts an international sporting competition.

Governments can gain by associating themselves with elite sports. In Australia, prime ministers sometimes attend sporting events, a mutually beneficial media opportunity, and by trying to associate themselves with sporting heroes. When hosting the Olympics, politicians take maximum advantage of the associated international prestige.

Finally, when identification with international sports teams is widespread, there are rewards for ordinary citizens: being an avid supporter of the team enables solidarity with friends and co-workers. If nearly everyone at the office is excited by an international match, then those who are uninterested are safer saying nothing—and supporting an opposing team can sometimes be awkward.

Tactics against alternatives
Alternatives to national identification in elite competitive sport—including abolition of international competitions, workers' games, and cooperative sports—are seldom mentioned. So it might be said that a key tactic against these alternatives is *cover-up,* except that so few people advocate such alternatives that active efforts to suppress information are hardly necessary.

International elite competitive sport has become hegemonic: it seems part of everyday reality. Those who are not interested in sports ignore the issue, and few of those interested in sports spend much time promoting alternatives. Meanwhile, young athletes see participation in a national team as an aspiration.

Then there are tactics to challenge nationalism in sports. This does not mean supporting a foreign team, because this doesn't question the importance of national identification of some kind. Let's consider some more frontal challenges. One is to denigrate international competitions, for example by exposing corruption, cheating and damaging side-effects. There is certainly plenty of critical material about the Olympics, for example exposés by journalists about the machinations of the International Olympic Committee.[8] Activists in host cities have tried to oppose the repressive and damaging measures used by governments to control the image portrayed about the games, for example moving homeless people out of urban areas and implementing harsh security measures.[9]

Every four years, a fresh crop of critics of the games emerges, especially in the host city. However, only a portion of their activity is directed against Olympics in general, or against the nationalistic dimensions of the games. Furthermore, in between Olympic years, there is little activity critical of the games or their patriotic dimensions. Possible tactics for challengers include exposing shortcomings and abuses (and plenty have been documented), denigrating the Games, explaining what is wrong with them, and mobilising protests.

There is one major obstacle to direct criticisms of any international sport: it is easy for others to say that this is

---

8 Andrew Jennings, *The New Lords of the Rings: Olympic Corruption and How to Buy Gold Medals* (London: Pocket Books, 1996).

9 Jules Boykoff, *Activism and the Olympics: Dissent at the Games in Vancouver and London* (New Brunswick, NJ: Rutgers University Press, 2014); Helen Jefferson Lenskyj, *Inside the Olympic Industry: Power, Politics, and Activism* (Albany, NY: State University of New York Press, 2000).

criticism of athletes. Elite athletes are sometimes treated as almost sacred: when they are unblemished in their personal and professional lives, they are considered beyond criticism. To question elite sport may be taken to imply, "You've been putting incredible effort into something that's not all that worthwhile."

Rather than mount a campaign against elite international sport, probably a better strategy is to promote alternatives—and there are many who do this. Increasing public participation in sport is a worthy alternative task, with well-documented benefits for health, personal satisfaction and social interaction. It would seem a reasonable step to argue that government expenditure should be redirected away from elite sport towards greater public participation. Cooperative sports—rather than competitive ones—are a complementary alternative, and might be promoted as a way of getting people to think of the disadvantages of competition.

Finally, there is another option: simply paying no attention to elite competitive sport, especially its nationalistic dimensions. Many people are already uninterested, but often they are polite about it. This could be encouraged, so that avid sports-watching is seen as uncool, or simply boring. This is already the case in some circles. Whether this could be the basis for something broader remains to be seen.

Conclusion

Sport can serve as a tool to promote nationalism. To do this effectively, participants need to be representatives of countries, so that engagement in the sport can be interpreted as a national enterprise. The sports need to be competitive, allowing individuals and teams representing countries to engage with those from other countries, so there is national honour involved in what would otherwise just be a contest between athletes. Ideally, the competitors are elite performers, without moral blemish, encouraging citizens to identify with the athletes representing their country. By glorifying national sporting heroes, especially winners, identification with one's country is encouraged, while governments bask in reflected prestige.

The role of these various components of international sport can be seen by imagining alternatives. A global fun run, in which participants are identified by some arbitrary characteristic such as birthday or height, would not provide much fodder for nationalism. A cooperative game, with participants joining for a common goal such as keeping a ball aloft, might foster a sense of international cooperation. A competition between non-elite performers—for example a swimming contest involving several presidents and prime ministers—would be more an amusement than a source of national identification, with internal opponents of any given president likely to support others.

Ironically, it is the seeming neutrality and non-political status of sport that makes it such a potent tool for national identification. Because sporting contests seem to be separate from politics and instead as places of moral virtue where the best athletes win, they are attractive to viewers, allowing them to identify with their preferred individuals or teams—and national identification comes as part of the package.

Because elite international competitive sports are so highly entrenched, it is difficult to challenge them. Direct criticism has a role, but perhaps more effective in the long run is promotion of alternatives, including mass participation in physical activity and cooperative sport.

6
Spying and surveillance

Spies: are they good guys or bad guys? The answer is easy: the spies on *our side* are good whereas the spies on *their side* are the worst of the worst.

Spying and surveillance are tricky for governments because of secrecy and obvious double standards. Let's look at some of the aspects and complications.

Spying on foreign enemies is the easiest case: it's assumed to be a good thing. However, to be effective, spying needs to be done covertly, so it's hard to praise spies in public. Furthermore, spying in general is often seen as a bit devious, so governments seldom boast that, "We have the best spies." Even mentioning the existence of current spies is a bit risky.

One solution is to praise past spying operations, done for a good cause. An example involves the Enigma machine, built in Britain during World War II to break Nazi secret codes. Breaking into codes is a type of spying, done at a distance, though it is perhaps better called surveillance. The story of the Enigma machine has been told in books and films, including the 2014 film *The Imitation Game*. It portrayed some British military figures unfavourably, with commanders being contemptuous of mathematicians and, after the war, showing serious bias against Alan Turing because he was gay. But this portrayal was in the overall context of the assumption that breaking German codes was a gallant, militarily crucial endeavour.

Very few people have personal experience of spying, or have even talked to a spy about what they do on the job. Consequently, ideas about spying are largely shaped by media coverage, much of it fictional in novels and films. In the widely read novels by John Le Carré, most of them set during the cold war, the world of spies is deceptive and morally challenging, with agents, double agents and double crossing. Overall the impression is that spying is somewhat disreputable. Indeed, spying requires lying, and thus has a taint about it.

Perhaps for this reason, as well as operational secrecy, governments say little about their own current spies. But when it comes to foreign spies, it is another matter: they are mightily condemned. (In practice, many foreign spies are monitored but never exposed; some are quietly expelled.) A few are arrested, tried and given long prison sentences, worse than if they had committed murder.

The most severe condemnation is reserved for insiders who serve the enemy: citizens, who are supposed to be loyal, who sell secrets or, even worse, reveal secrets because they believe in the cause of the enemy. Spying is cast into the mould of us versus them.

However, old-fashioned spying using agents has long been superseded by signals intelligence, which involves surveillance of electronic communications. All sorts of sophisticated techniques are used to monitor phone calls, emails and every form of electronic communication. Mostly this goes on in secrecy by all involved. Occasionally, though, there are stories about foreign dangers, for example hacking into databases by agents on behalf of North Korea or China. Because of secrecy, media stories are untrustworthy. Foreign governments seldom fess up saying "Yes, we were trying to access your vital data." Informed observers are wary: media stories may be due to strategic leaks intended to serve political objectives.

Some ways to refer to an agency

National security agency	This is the most serious-sounding terminology, implying grave responsibility. This is the most overtly state-oriented expression.
Intelligence organisation	The word "intelligence" has positive connotations because of the more common usages of the word, so this is a favoured expression by supporters of these organisations.
Surveillance operation	This emphasises a potentially negative side to agency activities.
Spy agency	This has negative connotations, given that spying is often seen as somewhat underhanded.
Secret police or political police	These terms highlight the capacity for political repression, and point to a connection with dictatorial regimes.
The spooks	This is an informal, humorous term.

A lot of surveillance is about economic information, for example trade secrets, designs and plans. Supposedly every government with suitable capacities does this, but it is usually kept secret. Occasionally there are popular cries to stop foreigners from "stealing our secrets," as though only foreigners engage in commercial espionage.

Then comes the most challenging surveillance of all: a government spying on its own citizens. In police states, this is a means of keeping control by monitoring dissent. In the former East Germany, the Stasi—the feared secret police—received information from one out of ten citizens in one of the most pervasive monitoring systems ever known. In the west, this sort of surveillance is condemned, so it is not surprising that western governments' own surveillance of their citizens is carried out in utmost secrecy.

Thinking in terms of in-groups and out-groups, there are two sets of processes going on here. Governments seek to build loyalty by encouraging citizens to think of themselves being part of a loyal in-group, and can foster this by creating, exaggerating or stigmatising out-groups. Foreign enemies are prime candidates for being out-groups and for bolstering in-group solidarity. Terrorists serve the same function, especially when they are seen as foreign or alien. But what if some of the "enemy" are actually part of "us"? This makes things trickier. The internal enemy could be communists, capitalists, ethnic groups, religious groups and so on. The risk to the government is that its own agents, including ones undertaking surveillance, will come to be seen as the enemy.

Consider the former Soviet Union, in which people were encouraged to report family members who were enemies of the state. For those who did this, one reward was greater identification with the state: for them, the out-group was class enemies. But for others, family loyalties were greater, and attempts by the government to encourage spying caused questioning of the state itself: for them, the state became an out-group.

Only in some circumstances can groups create loyalty that outweighs all competing loyalties. One of the reasons for the celibacy of priests in the Catholic Church is that it removes a competing source of loyalty: wives and children. Some cults require celibacy whereas others break down personal loyalties by expecting or mandating sexual relations with many different partners.[1] Governments have seldom been able to break down family loyalty; when they try, they risk being seen as the enemy of the people.

The governments of Australia, Britain, Canada, New Zealand and the US for decades had an intelligence-sharing arrangement called the Five Eyes agreement. Secret monitoring stations were set up to collect every possible electronic communication, and software developed to search the resulting data. This operation was so secret that its existence was hidden from the public, and even its name, Echelon, was secret.

New Zealand campaigner Nicky Hager made the first major breakthrough. Through conversations with workers at the facility at Waihopai run by the Government Communications Security Bureau, the New Zealand government's signals intelligence agency, he gradually

---

1 Lewis A. Coser, *Greedy Institutions: Patterns of Undivided Commitment* (New York: Free Press, 1974).

pieced together more and more information. The more information he obtained, the more he was able to suggest he knew more than he did, and thereby gather additional information. His 1996 book *Secret Power*[2] became well known among those who followed the machinations of government spy agencies, who also read James Bamford's *The Puzzle Palace* about the US National Security Agency and related exposés.[3] Hager's discoveries received some publicity when in the late 1990s repression-technology expert Steve Wright wrote about the Echelon surveillance system in a report to the European Parliament.[4]

Wider public awareness of massive western government surveillance of their own citizens did not occur until Edward Snowden's massive leak of documents from the US National Security Agency—the lynchpin agency in the Five Eyes agreement—hit the news in 2013.[5] Snowden's amazingly detailed information overshadowed previous

---

2 Nicky Hager, *Secret Power: New Zealand's Role in the International Spy Network* (Nelson, New Zealand: Craig Potton, 1996).

3 James Bamford, *The Puzzle Palace: A Report on America's Most Secret Agency* (Boston: Houghton Mifflin, 1982).

4 Steve Wright, "The Echelon trail: an illegal vision," *Surveillance & Society,* Vol. 3, Nos. 2/3, 2005, pp. 198–215.

5 For informative accounts, see Glenn Greenwald, *No Place to Hide: Edward Snowden, the NSA and the Surveillance State* (Hamish Hamilton 2014); Michael Gurnow, *The Edward Snowden Affair: Exposing the Politics and Media Behind the NSA Scandal* (Blue River Press, 2014); Luke Harding, *The Snowden Files: The Inside Story of the World's Most Wanted Man* (Guardian Books 2014).

findings, which were for the most part forgotten or ignored. The evidence was clear: massive government surveillance, carried out in supposedly democratic countries, was standard practice, not only against foreign enemies but also against ordinary citizens. It was bad when done by the East German Stasi. Why was it okay in the US?

Whereas previously the spying had been kept out of the public eye, not just for operational reasons but to prevent outrage, now it needed to be explained and justified. For governments and their apologists, a series of rationales emerged. One was to attack the messenger, calling Snowden a traitor. Another was to say, as had been said many times before, "If you've got nothing to hide, you have nothing to fear," implying that only criminals and terrorists should be concerned about government surveillance. There are many replies to this presumption in the form of a question. One of the easiest is to say, "In that case, please give me your credit card numbers and passwords."[6]

Governments can try to justify surveillance through the usual us-versus-them dichotomy, assuming surveillance is entirely against enemies of the state and people. The trouble is that many citizens start distrusting the state itself. This is apparent in the popularity of 9/11 conspiracy theories. Setting aside the question of whether President George W. Bush or other US officials actually had

---

6 Actually, the issues are more complicated than this. See Daniel J. Solove, *Nothing to Hide: The False Tradeoff between Privacy and Security* (New Haven, CT: Yale University Press, 2011).

anything to do with the planning or execution of the attacks on 11 September 2001, that so many people believe they might have suggests a deep-seated distrust of the US government.

Then there is the role of US spy agencies in other countries: they often team up with repressive govern- ments, in particular with security forces involved in surveillance, arrests, torture and killings. For example, rage in Egypt against President Hosni Mubarak, who stepped down in 2011 following massive protests, was in part directed against his ruthless security apparatus and, by association, US partners.[7] So there is an international dimension to outrage over spying on citizens: when governments share intelligence information against al- leged enemies, this can undermine trust among citizens who know about it.

Secrecy and surveillance

Scott Horton in his book *Lords of Secrecy* provides a powerful indictment of secrecy in US agencies involved in spying and surveillance. Horton argues that public discus- sion is essential for a democratic society, citing the example of ancient Athens, where citizens were involved in important decisions, including about security, namely going to war. Ancient Athens was successful in relation to its more authoritarian rivals, such as Sparta, because it was a "knowledge-based democracy," gaining strength from

---

7 Scott Horton, *Lords of Secrecy: The National Security Elite and America's Stealth Warfare* (New York: Nation Books, 2015), p. 157.

sharing and debating ideas from many individuals and sectors of society.

Horton traces the rise of excess secrecy in the US to the emergence after World War II of the national security elites, who dealt with nuclear weapons development and the challenge from the Soviet Union. He says the problem of unaccountable power was recognised by President Harry Truman and senior advisers who set up the Central Intelligence Agency; they established oversight mechanisms via the legislative branch of government, namely Congress. However, according to Horton, the huge size and resources of the spy agencies, combined with their use of secrecy, before long overwhelmed and captured their congressional overseers. Secrecy became a tool to build bureaucratic empires, to hide failures and to carry out policies without scrutiny.

The next sector of society with the potential to restrain the agencies was the media, but the US mass media became tools of the state, being reluctant to break stories about any sort of abuse, for the example the 1968 My Lai massacre in Vietnam or the torture at Abu Ghraib prison revealed in 2004. So, according to Horton, the one remaining group with the potential to challenge unaccountable secrecy is whistleblowers, who have become a target for suppression.

Horton's analysis points to the powerful role of secrecy in agencies involved in spying and in undeclared war, in particular the use of drones for extra-judicial assassination. Secrecy can become an end in itself. Horton himself is not making an argument against surveillance or drones or wars. He just wants there to be an open discussion so that better informed decisions, with support from politicians and the public, can be made.

This is enough background to indicate the complexities of spying and surveillance in relation to building loyalty to the state. Basically, the government has to pursue seemingly contradictory directions, maintaining secrecy for operational reasons and to hide corruption and abuses, while somehow convincing members of the public that monitoring them is for their benefit.

In the following sections, I first outline tactics to build loyalty to the state in relation to spying and surveillance, then tactics against alternatives to the standard approach, and finally tactics to challenge surveillance.

Tactics to build loyalty

The first tactic is *exposure* of good things about the state. Here the challenge is the greatest. The safest approach is to expose only achievements, such as spying successes in past wars and successes in preventing terrorism. However, this has to be done carefully so as to suggest that bad guys are the only targets. By carefully picking stories to release, and angles on those stories, the aim is to encourage people to *value* the role of intelligence services, positioning them as protectors of the population.

Their role is *explained* as a necessary function of maintaining security. Part of the explanation involves suitable framing. Rather than refer to spying and surveillance, the usual language is of intelligence and national security.

Governments routinely *endorse* their intelligence agencies, and *reward* them generously with good salaries and conditions, as part of ample budgets that signify the importance of their task.

In these ways, governments try to build citizen loyalty to the agents of control. However, compared to many other areas—museums, elections, sport, education, media—the task is greater because spying itself is often seen as a shady sort of activity, involving deception and underhanded methods. It's a bad method of achieving a good goal, and the negative associations with the method tend to rub off on the goal. So for many governments, the less said the better. Justifications are only brought forth when the issue has been publicised or when arguing for greater resources. Their ideal technique is to condemn spying by other governments and hope that no one even thinks about their own spying.

Marginalising alternatives
Are there any alternatives to the usual government spying? This is a difficult question to answer, because there is so little discussion of alternatives. Let's consider some possibilities.

One alternative is to say there should be no spying at all. This is easy to challenge, because the bad guys—foreign governments—are spying on us, so we need to spy on them. So the no-spying option is usually posed as, "There should be no spying on our own people." This is actually a radical alternative in countries where the government is repressive and nearly all surveillance is against internal opponents. To this option, governments regularly use the method of fear-mongering, raising the alarm about terrorists, communists, traitors, heretics or others who threaten the fabric of society, in other words the government.

There are, in some cases, actual opponents who pose some danger to the public: terrorists and criminals for example. Such opponents are valuable for governments because they help justify spying on everyone. For the moment, assume there are legitimate reasons for surveillance. How should it be done?

The usual approach is to have a system but make sure it is under legitimate political control, for example with scrutiny by elected politicians, who supposedly serve as agents of the public. The trouble is that spy agencies become too powerful and can win over their political masters, invoking the necessity of secrecy to ensure that effective controls are seldom invoked. On a more nasty level, spy agencies can collect dirt on politicians, implicitly threatening to covertly release the information. The FBI under J. Edgar Hoover supposedly engaged in this sort of blackmail. It is the sort of technique used by criminal organisations: demand participation in crime and then use the possibility of exposure to deter disloyalty.

So what about alternatives that involve something completely different? One possibility is promoting social justice. Rather than spying on opponents, instead address the sources of their grievances. This is good for a long-term view, but does not address the possibility of immediate threats.

One alternative is to introduce a "citizens inspectorate," namely citizens who have the power to check what spy agencies are doing and to make reports and recommendations. To be effective, a citizen inspectorate would need to be sizeable and have a significant turnover to prevent capture by the agencies.

Some agencies already have an oversight body or individual, for example an inspector-general to whom complaints can be made by employees or members of the public. The trouble with such systems is that they usually become closely aligned with the agency, the same problem that occurs with legislative oversight.

If citizen inspectors were randomly chosen and served short terms, they would be less likely to be able to bought off or intimidated: some of them might be independent enough to make probing assessments and discourage abuses.

Agency heads would detest such a proposal, no doubt arguing that citizen inspectors, lacking security clearances, could not be allowed to know what agencies are doing. This objection is the familiar claim that secrecy prevents scrutiny.

Another alternative would be to set up a secure avenue for leaks from agencies. By analogy with WikiLeaks, it might be called SpyLeaks. This would enable abuses to be exposed with less likelihood of reprisals. Then comes the question of who would have access to the leaks. Perhaps legislators, or citizen inspectors, or even the general public.

Given the efforts of the US government to shut down WikiLeaks, it is obvious that SpyLeaks would never get off the ground. If it were ever implemented by agencies themselves, it might well have a back door so that agency officials could identify the leakers.

Giliam de Valk and I wrote an article about "publicly shared intelligence."[8] Giliam in his PhD research compared the performance of the Dutch intelligence services, which operated with the usual secrecy, with a very different sort of intelligence operation: the Shipping Research Bureau. The Bureau operated at the time of apartheid in South Africa, when there was an international embargo of oil imports as a form of pressure against the regime. However, some companies broke the embargo, sending their ships surreptitiously to deliver oil to South Africa. The Bureau sought to collect information about these rogue traders and expose them, thereby shaming the companies.

The Bureau used secrecy in some aspects of its collection and analysis of data. Individuals sent the Bureau information about ships, and it sought to verify this information, but did not release the names of its informants. But the Bureau's reports were public. Unlike spy agencies, it made its assessments available for scrutiny.

Giliam in his research found that the Bureau's reports were far more accurate than reports of the Dutch intelligence agencies. Publicly shared intelligence apparently had an advantage. This was what you might expect: open scrutiny improves quality. The same thing happens in science. The quality of the open scientific literature, which is subject to peer review before publication and available for scrutiny by anyone after publication, is widely

---

8 Giliam de Valk and Brian Martin, "Publicly shared intelligence," *First Monday: Peer-reviewed Journal on the Internet,* Vol. 11, No. 9, September 2006.

regarded as superior to secret corporate or government research. Similarly, open source software, in which the code is publicly available for scrutiny, is usually superior to proprietary software.

Publicly shared intelligence thus offers an alternative to the usual government surveillance. By drawing on the resources of the entire population both for inputs and evaluation of assessments, this form of intelligence would have the advantages of open source alternatives. (We didn't call it open source intelligence because that name was already used for a different alternative: intelligence drawing on openly accessible information, but lacking the open scrutiny essential for quality control.)

Publicly shared intelligence would be a frontal challenge to conventional intelligence operations built around secrecy. As expected, there has been no government interest in this alternative. For all practical purposes, it is invisible. No government has sought to test it.

From this brief discussion of ways to provide stronger oversight of spy agencies, it should be obvious that agencies will do nothing to publicise options that enable significant independent citizen involvement, much less actually implement them.

Challenging government surveillance
A key method of challenging surveillance is to expose it. Secrecy serves spy agencies by hiding abuses and failures. The bigger the abuse, usually the greater the secrecy.

Whistleblowers, leakers, investigators and journalists play crucial roles. Edward Snowden revealed unparalleled amounts of inside information. He was highly effective because he kept a low profile until he had gathered the information. (He kept his plans secret.) He then carefully chose a journalist and media outlet—Glenn Greenwald of the *Guardian*—to whom to release the information. When Greenwald wasn't responsive, Snowden contacted Laura Poitras, a dissident filmmaker and friend of Greenwald's, and arranged to meet them. Snowden chose well: the *Guardian*'s editors refused to buckle to pressures from the National Security Agency and its British equivalent, and went ahead with exposé after exposé.

Another exposure technique is to reveal the identities and activities of spies. The magazine *CovertAction Information Bulletin* beginning in 1978 published the names of a number of CIA agents. So effective was this outing that in 1982 the US Congress passed a law making such disclosures illegal and subject to severe penalties. This response suggests the power of exposure: spies aim to gather information about others but they don't want information gathered about themselves: their efforts rely on secrecy and deception, for example false identities.

Today, it is far easier to collect and publish information. Citizens with digital cameras can record police use of force as it happens, in many cases exposing abuses that in previous decades would have been hidden from the public. Similarly, recording of the identities and activities of spies can be a powerful technique.

Another important technique is to counter the justifications for surveillance. This is a big area. One technique used by agencies is to lie about the value of information gathered, for example in preventing terrorist attacks. Critics can expose the failures of agencies, for example in not picking up on clues about the 9/11 attacks or not anticipating the Arab spring. There were important failures decades ago too, for example the falsity of the alleged "missile gap" between the US and Soviet nuclear arsenals in the late 1950s, and the failure to anticipate the collapse of communist regimes in 1989. These were all failures of US agencies; there would be equivalent shortcomings in agencies in other countries that need to be exposed and criticised.

Next is the issue of official channels. Many governments establish laws and regulators for privacy protection. In practice, though, these seldom do much to control surveillance operations. Indeed, there is a body of writing on how privacy protection is routinely outflanked by technological developments and rogue operations.[9] What does privacy legislation do in the face of ever-expanding use of security cameras? What about revenge porn, when people post sexual images of former sexual partners? What about the Five Eyes surveillance of citizens?

Most employees tasked with enforcing privacy laws and regulations do their best, and no doubt many worthwhile protections have been implemented. But this is a losing effort in the face of an onslaught of monitoring capacities, including ones where people voluntarily offer information that potentially can be used against them, mostly in social media, also subject to monitoring and analysis by governments.

---

9 For example, Simon Davies, *Monitor: Extinguishing Privacy on the Information Superhighway* (Sydney: Pan Macmillan, 1996), chapter 6.

Rather than rely on privacy protection to limit surveillance, a more promising approach is to mobilise support, indeed to build a social movement. But despite people's serious concerns about government surveillance and many abuses, there is little sign of the development of a broad-based anti-surveillance movement.

There are many initiatives. The group Anonymous has taken direct action online in support of WikiLeaks. There are many supporters and users of encryption who oppose efforts by US government officials to mandate backdoors to encryption systems using the rationale of needing to be able to track down terrorists. Then there are software developers and entrepreneurs making accessible the means to avoid surveillance. These include the developers and promoters of the Tor browser, search engines like duckduckgo that do not record searches, convenient encryption systems and anonymous remailers, among others. A basic test is to ask, "Would this system be useful to dissidents in a repressive regime?" If it is, then it is probably worth promoting everywhere, including in countries where governments ostensibly respect civil liberties, because when it comes to surveillance, lots of governments are seeking powers that can easily be used to suppress dissent—and quite possibly are, given the secrecy involved in the whole system.

Part of challenging surveillance is resisting it, and that is not easy in a world with ubiquitous monitoring. It's possible to keep a low profile, but this might involve considerable inconvenience, for example not having a credit card, not driving (in areas where vehicle licence numbers are monitored) and not using a mobile phone.

Another form of resistance is to insert incorrect information into databases, for example "accidentally" using a slightly different birthday or address for different databases, or perhaps some politician's phone number. Although this can make it more difficult to collate data about you—you may end up with lots of nearly identical but slightly different versions of yourself on databases—it does little about surveillance more generally. Fake profiles on Facebook, Google and other platforms are common, many of them manufactured and sold to enhance the buyer's online image.

Because remaining outside routine surveillance is so difficult, and putting false information into databases usually has a marginal impact, probably a better form of resistance is to make public statements or otherwise protest surveillance openly. Some opponents set out to disable security cameras. Others perform colourful protests in front of the cameras for the delectation of operators.

Spying and patriotism revisited

There are various ways to oppose spying operations, but how do these relate to state power? To start, much surveillance is undertaken by the state, so opposition directly challenges state power. Other surveillance is undertaken by companies, for commercial purposes. Facebook and Google collect information about users to better direct advertisements, the lifeblood of their operations. However, as Snowden's leaks revealed, spy agencies use various means to tap into private information streams.

Probably just as importantly, private data collection makes people become used to exposing their lives online, without thinking about how data is being collected by banks, phone companies and social media companies. Surveillance is increasingly seen as normal, as nothing much to worry about. When people regularly reveal details about their lives to anonymous companies and government agencies, they are likely to come up with rationalisations to justify what they do. This helps explain why anti-surveillance has not become a major social movement.

However, governments are still caught in their own contradictions. They undertake surveillance, but want to keep it secret and therefore have difficulty justifying it when it is exposed. They want to make people believe that all spying is on bad guys, but then are exposed spying on their own citizens. So they point to the dangers of criminals and terrorists, but at the risk of becoming tainted by their association with internal spying, often associated with repressive regimes.

Government thus can have a hard time finding the optimal balance between hiding and justifying their spying operations. Surveillance is not a good means for them to drum up support. Opponents can use the inherent contradictions in state surveillance in mobilising resistance, but have their own challenges in trying to get people to care enough to act, given the gradual encroachment of data-gathering methods and the immediate benefits to individuals in acquiescing to this data-gathering.

Perhaps the most powerful technique is to use the expanded capacities for collecting data against government agencies themselves. Already, police are changing their behaviour because of the ubiquity of cameras recording their actions. Perhaps government officials may decide to change their operations if they start becoming the target of citizen surveillance.

7
Terrorism

On 15 December 2014, a man named Man Haron Monis took hostage a group of patrons at the Lindt café in Martin Place, in downtown Sydney. The police Tactical Response Group was called. There was a stand-off lasting over 16 hours. In the dramatic climax of the siege, Monis killed one of the hostages, the police stormed the café, another hostage was killed (probably by a stray police bullet) and so was Monis.

This event received saturation coverage in the media, with continuous television treatments and page upon page in the daily newspapers. After the siege was over, there was an outpouring of sympathy for the two hostages who died, with Martin Place being covered with thousands of bouquets.

The siege seemed to unite people in support of the state.[1] The prime minister, Tony Abbott, took a strong stand against Monis' action and in support of the police, and the federal opposition leader, Bill Shorten, backed him to the hilt.

---

1 Paul H. Weaver, *News and the Culture of Lying* (New York: Free Press, 1994), makes the point that news is oriented to crisis, thereby promoting crisis government, giving greater power to the executive and removing power from routine decision-making processes.

Was it a terrorist incident? This was debated in the aftermath. Monis certainly was not a typical terrorist, and was not part of any group making demands. The most common view was that he was a "disturbed" individual, with a long history of crimes and strange behaviour.

Association with Monis was toxic politically. Some years earlier, the New South Wales opposition leader, John Robertson, had written a letter in support of Monis, who was a constituent. Although this was nothing special at the time, after the siege it was deemed sufficient to trigger a push for Robertson to resign.

Whether or not Monis' siege counts as terrorism, it served much the same function—from the point of view of the state. It illustrates how terrorism serves the state. US President George W. Bush, in the aftermath of the 11 September 2001 terrorist attacks, declared, "You are either with us or with them [the terrorists]."[2]

The state is normally considered to include the government, various government agencies, and perhaps government-owned businesses. The eminent sociologist Max Weber defined the state as the governing entity claiming a monopoly over the use of legitimate violence— legitimate in the eyes of the state. "Legitimate violence" here refers to the police and military. Armed challenges to the state are considered illegitimate, and are to be repressed without reservations.

The basis for the legitimacy of the state is that it protects the population against threats, most dramatically the threat of invasion, conquest and subjugation. In times

---

2 See also the discussion of this quote in chapter 8 on language.

of war, the power of the state increases dramatically in order to defend the population—and the state itself.

Terrorism provides a substitute for war in terms of mobilising support for the state. Citizens identify with the government and look to it for protection. If "War is the health of the state," terrorism is a booster shot.[3]

Why is terrorism so effective in boosting state power? After all, many people die every day, for various reasons. Some die from disease; some are killed in traffic accidents; some are murdered; some kill themselves. Furthermore, in most places these and other dangers cause far more deaths than terrorism. In many countries, traffic accidents kill hundreds or thousands of people per year, and many could be prevented by safer roads or by diverting travellers to safer modes of transport, such as trains. After 9/11, many US travellers avoided planes and drove instead. Because driving is much riskier than flying, the death rate from travelling accidents increased, perhaps raising the death toll by more than the 9/11 attacks themselves.[4]

It is worthwhile, therefore, looking at the mechanisms by which terrorism serves to generate support for the state.[5] The first tactic is exposure. A siege in a café,

---

3 Randolph Bourne famously said, "War is the health of the state." See chapter 13.

4 Gerd Gigerenzer, "Dread risk, September 11, and fatal traffic accidents," *Psychological Science,* Vol. 15, No. 4, 2003, pp. 286–287.

5 The exposition here presents the system-support tactics outlined in chapter 1.

with hostages, is ideal fodder for media coverage. It has drama, danger and an enemy, with police as the saviours, providing a story that combines fear and potential reassurance. Traffic accidents and heart attacks seldom offer such a compelling narrative.

In large part, terrorism obtains media coverage because it is designed to do so. Some analysts have described terrorism as "communication amplified by violence."[6] The goal of what is conventionally called terrorism is to capture public attention. The victims of the terrorists are not the actual targets, but tools to generate attention. The media come calling and provide the conduit for gaining awareness from the wider public.

Terrorist attacks provide an ideal opportunity for agents of the state—police or the military—to be heroes. They respond to the threat, becoming the protectors of the population. In this way, protection of the state becomes fused with protection of the population. The state is seen as the guardian of public safety. Terrorists are cast as villains, as pure evil. For the purposes of the state, the terrorists need to be evil, so a classic morality play is enacted. Humanising the terrorists—seeing them as

---

6 Alex P. Schmid and Janny de Graaf, *Violence as Communication: Insurgent Terrorism and the Western News Media* (London: Sage, 1982). See also Brigitte L. Nacos, *Mass-Mediated Terrorism: The Central Role of the Media in Terrorism and Counterterrorism* (Lanham, MD: Rowman & Littlefield, 2002); Joseph S. Tuman, *Communicating Terror: The Rhetorical Dimensions of Terrorism* (Thousand Oaks, CA: Sage, 2003).

regular people, perhaps even fighting for their ideals—
would confuse the message.

Terrorism is usually explained to the population in
simple terms: the bad guys, the terrorists, are trying to
harm "us" and destroy "our" way of life.[7] Other factors are
ignored or skated over, such as the harm or injustice that
might have created grievances (especially harm done by
the state itself), the double standards involved in ignoring
state terrorism (discussed later), or that there might be
better ways to deter or discredit terrorism. Official
explanations for terrorism almost never mention that if
suitable opportunities for citizens to express their views
existed, many grievances would evaporate. In cases of so-
called "international terrorism," almost always there are
"international grievances"—government involvement in
foreign countries, such as invasions, occupations, corporate exploitation or drone attacks—for which no opportunities for citizen participation in decision-making exist.

The most important technique by which terrorism is
interpreted by the state is framing, usually in a Hollywood
template with the government as the good guys and the
terrorists as the bad guys, with the only way for the good
guys to win being through superior force. With this way of
thinking, terrorism provides an unquestionable justification for state violence.

Anti-terrorism is enshrined through laws and regulations. In this way, the state indicates that terrorists are the
official enemy, and that opposing terrorism is legally

---

7 Ziauddin Sardar and Merryl Wyn Davies, *Why Do People Hate
America?* (Cambridge: Icon, 2002).

mandated. Indeed, anyone who does not go along with this agenda might be caught up in anti-terrorism laws and regulations. The connection between anti-terrorism laws and patriotism is most obvious in the US Patriot Act, an anti-terrorism law passed after 9/11. The acronym[8] is intended to indicate that anti-terrorism is patriotic.

The state's agencies usually give a stamp of approval for anti-terrorism policies, with the main debates occurring within a narrow band of disagreement of how unrestrained agencies can be. A whole range of agencies may be involved: government executives, parliaments, courts, the military, police, spy agencies, and corporate contractors. By going along with government anti-terrorism agendas, they help legitimise them.

Finally, anti-terrorism is imposed on the population through repressive measures, including extensive surveillance, interrogations, arrests and show trials. Vocally opposing the government's anti-terrorism agenda may be enough to trigger targeted surveillance, harassment (for example, extra screening at airports), denial of jobs, or worse. Imposing penalties, formal or informal, for being critical of anti-terrorism discourages dissent. On the other hand, those who enthusiastically join in the anti-terrorism chorus may be rewarded with jobs, promotions, research funding and media opportunities. Conspicuous patriotism, via anti-terrorism, can pay.

---

8 The USA PATRIOT Act stands for Uniting and Strengthening America by Providing Appropriate Tools Required to Intercept and Obstruct Terrorism Act.

Thus in a range of ways, governments can mobilise support by drumming up concern about terrorism. The irony is that terrorists play right into the government's hands.

Terrorism backfire

A physical attack on civilians is a powerful method of gaining attention. As noted earlier, it is a mode of communication, using violence against civilians to send a message to a broad audience, with special salience for governments.

Normally, when groups do something seen as unfair, or just bad, they try to reduce public outrage by hiding their actions, disparaging the targets, explaining away their actions, using official channels to give a stamp of approval, and intimidating or rewarding people involved. Although harming innocent civilians is widely seen as reprehensible, do terrorists use any of these methods to reduce outrage? Quite the contrary: terrorists routinely try to *increase* outrage.[9]

The most powerful terrorist actions are open rather than hidden. Bombings or shootings are done in public. Sometimes terrorists film and publicise their atrocities, for example beheadings. They often try to maximise media coverage. The 9/11 attacks were highly successful, occurring in broad daylight for all to see, targeting icons of US capitalism and the state. Individual terrorists may try to

---

9 Many of the ideas here are addressed in Brian Martin, *Justice Ignited: The Dynamics of Backfire* (Lanham, MD: Rowman & Littlefield, 2007), chapter 12.

hide their identity, but usually their organisations take responsibility for acts. That is the whole point: terrorists are trying to gain attention through the use of violence.

Terrorists can do little to reduce public outrage from their acts. They have minimal capacity to devalue their targets or to use official channels to give an appearance of justice. They seldom have access to sympathetic media to reinterpret their acts by lying about what they have done, blaming others, or minimising the consequences. Indeed, they are just as likely to exaggerate the impact.

So it seems that terrorists do everything possible to generate outrage over their actions. They almost seem to want to make violence backfire against them, generating greater disgust and opposition. How then can terrorism be considered a rational strategy? The one plausible explanation is that terrorists hope their opponents, who are much stronger, will over-react, use excessive state violence and trigger greater resistance to the government. Other explanations involve processes that are less functional for achieving the explicit goals of the terrorists. Terrorism can be an expression of resentment, getting back at detested governments or officials. It can build in-group solidarity, and attract new followers, through a type of initiation, but at the expense of generating greater opposition at the same time. Most terrorist acts are carried out by men; using violence can be a way of asserting male superiority and excluding most women.

Whatever the reasons, anti-state terrorism serves the state, so there is a mutually reinforcing interaction between states and their violent opponents, with neither side having much incentive to search for alternatives. Yet, if terrorism is considered purely in functional terms, namely being effective in achieving its goals, then nonviolent alternatives would be far superior in most cases. But for states, terrorists provide the ideal opponents, offering a rationale for their own violence.

The words "terrorism" and "terrorist" are widely used as if they have a clear meaning. I have used them here to refer to the use of violence against civilians by non-state groups, with al Qaeda's 9/11 attacks as a prime example. However, looking more closely at the concept of terrorism soon generates confusion.[10] There are actually dozens of different definitions. Furthermore, governments seldom bother with academic definitions, but simply label their opponents terrorists. The US government, fighting in Vietnam, labelled the National Liberation Front, commonly called the Viet Cong, as terrorists. In South Africa under the racist system of apartheid, the government labelled its opponents, the African National Congress, as terrorists. In the Philippines, the government labels its armed opponents, engaged in a rebellion in rural areas, as terrorists. In India, Maoist rebels fight the government in parts of the country; the government calls them terrorists. But in these conflicts, governments often

---

10 See Conor Geerty, *The Future of Terrorism* (London: Phoenix, 1997) for a critique of the expression "terrorism" as originally referring to state terror and eventually becoming an incoherent term of condemnation. On the peculiar logic underpinning anti-terrorist practices, see Richard Jackson, "The epistemological crisis of counterterrorism," *Critical Studies on Terrorism*, Vol. 8, No. 1, 2015, pp. 33–54.

are responsible for far more rape, pillage, torture and murder than their opponents. So perhaps these governments should be called terrorists too.

That is exactly what some scholars have done. They take the term "terrorism" at its face value, namely as referring to actions that strike terror into the minds of citizens, and note that by this definition, governments are by far the biggest terrorists. High-level aerial bombing can be just as terrifying as explosions in marketplaces, and torture by governments can be just as devastating as torture by insurgents. Terrorism by governments is called "state terrorism."[11]

In the Indochina war, two or three million Vietnamese, Cambodians, Laotians and others died due to US military actions, which included bombing, torture, assassinations (tens of thousands of them), and forced movements of populations into secure compounds, which might be called concentration camps. A large percentage of the victims were civilians. Similarly, in places like Guatemala

---

11 Noam Chomsky and Edward S. Herman, *The Political Economy of Human Rights* (Boston: South End Press, 1979); Frederick H. Gareau, *State Terrorism and the United States: From Counterinsurgency to the War on Terrorism* (Atlanta, GA: Clarity Press, 2004); Alexander George (ed.), *Western State Terrorism* (Cambridge: Polity Press, 1991); Michael Stohl and George A. Lopez (eds.), *The State as Terrorist: The Dynamics of Governmental Violence and Repression* (Westport, CT: Greenwood, 1984); Michael Stohl and George A. Lopez (eds.), *Terrible Beyond Endurance? The Foreign Policy of State Terrorism* (Westport, CT: Greenwood, 1988). See also the discussion of state crime in chapter 4.

and Indonesia, where hundreds of thousands of civilians have been killed, nearly all the killing has been on behalf of governments.

When governments undertake large-scale killing, they nearly always accompany this by measures to reduce public outrage.[12] They usually

- hide what they are doing, at least from wider audiences
- devalue their targets (using the label "terrorists" is just one technique)
- reinterpret their actions by lying (for example, civilians killed are called insurgents), minimising consequences, blaming others (such as "rogue ele-ments" being covertly funded) and framing their actions as worthy (for example, protecting national security)
- use official channels to give an appearance of justice (such as formal inquiries into killings)
- intimidate and reward people involved, including journalists and witnesses.

The double standard is stark.[13] Governments kill, or threaten to kill, large numbers of civilians, something that

---

12 Brian Martin, "Managing outrage over genocide: case study Rwanda," *Global Change, Peace & Security,* Vol. 21, No. 3, 2009, pp. 275–290; Brian Martin, "Euthanasia tactics: patterns of injustice and outrage," *SpringerPlus,* Vol. 2, No. 256, 6 June 2013, http://www.springerplus.com/content/2/1/256.

strikes terror into the hearts of potential victims. Yet many of these same governments are able to escape censure for their own activities, while pointing the finger at allegedly dangerous enemies, the so-called terrorists, turning their comparatively low-level attacks into justification for massive mobilisation and retaliation. This double standard is accomplished by parallel sets of tactics, on the one hand to reduce outrage from the government's own actions and on the other to mobilise outrage against the "terrorists."

It is not surprising that there is vastly more scholarship on non-state terrorism than on state terrorism, and that the very idea of state terrorism is almost never presented in the media or textbooks and is largely unknown to the wider public. It is in this context that it is possible to say that terrorism strengthens the state. This doesn't happen automatically: governments do everything possible to ensure that it does.

In the face of armed opposition, governments might adopt measures to de-escalate conflict, for example by promoting social justice, opening avenues for citizen participation, prosecuting government agents involved in torture and killing, and introducing a range of measures to promote reconciliation. In a free and open society, with opportunities to bring about change through the system, terrorism would lose much of its attraction, and it would not aid recruitment or popular support.

---

13 See also Brian Martin, "How activists can challenge double standards," *Interface: A Journal for and about Social Movements,* Vol. 7, No. 2, 2015, pp. 201–213.

What often happens instead is an insidious process of reinforcement. After an anti-state terrorist attack, the government responds massively, for example with arrests, torture or bombings—and in the course of this response harms previously uninvolved civilians. This results in new grievances, giving support to insurgent groups, who mount further attacks, leading to more reprisals, and so forth. The government, by choosing repression as its response to terrorism, fosters the very conditions that stimulate more terrorism. Do governments seem to worry about this? In many cases, not at all. The more they are attacked, the more governments gain greater power and legitimacy.

This pattern was apparent in Afghanistan after the western invasion in October 2001, supposedly in retaliation for the 9 September 2001 attacks in the US. (Nearly all the 9/11 attackers were from Saudi Arabia.) Bombing in Afghanistan killed thousands of civilians, but this was not publicised in the west, a type of cover-up. The intended targets, the Taliban, were demonised as terrorists, even though the CIA had supported them in the 1980s after the Soviet government invaded Afghanistan. The bombing of Afghanistan was explained as part of the war on terror, even though it terrorised the Afghani population. The attack was authorised by the United Nations Security Council some time afterwards.

If anyone wants to increase the power of the state, a terrorist attack is probably the single most effective way to do so. After 9/11, there was enormous international sympathy for the US government and people. The government massively increased military funding and especially funding for national security. Dissent was portrayed as a threat. Patriotism was given an enormous booster shot. The same thing has happened in other countries after terrorist attacks, including Australia. In October 2002, there was a bombing in Bali; though this was in Indonesia, the primary victims were western tourists, with 202 killed, 88 of them from Australia. The number of Australians killed was nearly as high a proportion of the Australian population as the 9/11 death toll was of the US population. Similarly, legislation was introduced to give much more power to security agencies, and their funding was increased dramatically.

What to do?
For those who are critical of excessive patriotism and wary of the power of the state, what can be done to oppose the role of terrorism in strengthening the state? This is a very big subject, so only a few possible actions and initiatives can be mentioned.

On an individual level, it is possible to become better informed about violence around the world, to be better able to put terrorism in context. Since the end of the cold war, there have been dozens of major conflicts, with the most deadly ones being in Africa, including the Congo, Algeria, Rwanda, Sudan and Burundi: in each of these countries, hundreds of the thousands of people have died in wars or genocides. The wars in the Congo have been the most deadly, with some five million deaths. Compared to this, international terrorism leads to relatively few deaths. The implication is that the threat from non-state terrorism in the west has been blown out of all proportion—thus serving to strengthen states—while more serious threats to the lives and safety of the world's population are mostly unknown to wider audiences. Becoming aware of the figures and examples can provide an antidote to the continual drum-roll about dangers from terrorism.[14]

It is also worth studying the figures about other threats to personal safety, such as traffic accidents, drowning in bathtubs, falling over and domestic violence. For most people, these are much greater threats to safety than terrorism.

Another approach is to support alternatives that undermine the attractions of terrorism for potential terrorists. Greater social justice—treating people more fairly, and addressing grievances—can foster commitment to a society. Also important is opening channels for change through the system. When people feel that they are being treated badly and that there is no legitimate way to make a difference, some of them may want to resort to violence, even when it is counterproductive.

Research shows that methods of nonviolent action, such as rallies, strikes, boycotts and sit-ins, are usually more effective than violence in achieving the goals of campaigners. Spreading the message about the power of nonviolent action, and developing campaigns that use this power, provide models for others to follow.[15]

---

14 Virgil Hawkins, *Stealth Conflicts: How the World's Worst Violence Is Ignored* (Aldershot, UK: Ashgate, 2008).

15 For specific applications to terrorism, see Tom H. Hastings, *Nonviolent Responses to Terrorism* (Jefferson, NC: McFarland, 2004); Senthil Ram and Ralph Summy (eds.), *Nonviolence: An*

Although nonviolent action may be more effective, the sad reality is that governments seldom promote it, but rather raise the alarm about terrorism, repress dissent, resist nonviolent protest, and create the conditions that foster terrorism. Nonviolent campaigners thus face a double challenge: to demonstrate to others that nonviolence is a better option than violence, and to confront authorities that resist peaceful change and thus create conditions that stimulate violence. This is the challenge of dealing with a government-terrorism symbiosis.

When alarms about terrorism are raised, another approach, at an individual level, is to say "ho, hum" and treat the whole issue as unimportant. Whenever terrorism is reported on television, change the channel. If everyone ignored it, the purveyors of concern about terrorism would lose credibility. Unfortunately, this approach would not make much difference unless adopted by a large number of people.

Humour is another response. Indeed, quite a few people feel that terrorism alarms are silly, and make jokes about them. This can be risky at airports, where authorities over-react to the slightest comment. Some types of humour may be safer and more revealing. A "supportive" humorous political stunt involves pretending you support the cause you are making fun of. For example, you could go around an airport or railway station reporting

---

Alternative for Defeating Global Terror(ism) (New York: Nova Science Publishers, 2008). For my approach, see "Nonviolence versus terrorism," *Social Alternatives,* Vol. 21, No. 2, Autumn 2002, pp. 6–9. See also the discussion in chapter 13.

unattended bags—even if unattended only briefly—or perhaps reporting "suspicious behaviour" by well-dressed businessmen. The next step is to work in teams. One member leaves shopping bags unattended, each one containing a balloon, or a present for the finder, while another reports these potentially dangerous bags to the authorities. However, stunts like this could go seriously wrong if there was an actual attack while staff were investigating false alarms.

My assessment is that it is not easy to develop a campaign to address the out-of-proportion alarm about terrorism. Governments do what they can to tout the risk, and this feeds perfectly into media news values, while meanwhile more serious problems are neglected. At a basic level, the first step is not to get caught up in the terrorism alarm, but beyond this, it is difficult to develop a campaign to change the agenda. This is an area where social experimentation is needed: activists can try out various ways to redirecting attention, making fun of terrorism alerts, promoting non-state responses, or in other ways addressing the mutual reinforcement cycle between states and terrorists.

8
Language

"We invaded Iraq." I've read this statement numerous times. It refers to the 2003 invasion of Iraq, but the times I've seen it, the author is not a US soldier, commander or policy-maker, but instead a critic of the invasion. These US critics are disgusted by the lies and damaging actions of the US government—their own government! Hence the word "we."

Critics know full well the invasion was decided upon by George W. Bush, Dick Cheney, Donald Rumsfeld and company, sold to a few other governments and carried out through military chains of command. To say "We invaded" is shorthand for something like "Top decision makers in the US government ordered the US military to organise an invasion. Isn't it terrible that 'our' government did this?"[1]

The trouble with "We invaded Iraq" is that it collapses the distinction between the government and the population. "We" suggests that the writer identifies with the government.

A US government official who supported the invasion of Iraq would never say, "We protested against the invasion," meaning that people in the US protested. Pro-

---

1 The word "our" only works for US readers. Foreigners cannot be expected to feel ownership of or association with the US government.

testers are different from, indeed against, the government: protesters are "they."

The uses of "we" and "they" in relation to the invasion of Iraq provide an example of how assumptions about people and governments enter language and then are strengthened in people's minds by the constant repetition of that language. This is a very big topic, and I'm only going to touch the surface by mentioning several examples in which language reflects and promotes the identification between individuals and the state.

Consider these different entities:

• Country: a geographical area, encompassing people, institutions and much else
• Government: the system of political leaders or rulers
• People: everyone living in a country

In most news reporting about national and international affairs, the country, government and people are not distinguished. Think of "Berlin today said," "The US intervened" or "Britain is reluctant." In media conventions applying to international affairs, the name of the country or the capital city is treated as referring to the government or, more precisely, top officials in the government.

The effect of this sort of language is that it is difficult to talk about—and think about—situations in which people's views or actions differ from those of government policy-makers. Let's go back to the 2003 invasion of Iraq. Shortly before the invasion, there were massive rallies across the world, the largest anti-war protest in history.

Millions of people demonstrated their opposition to the impending war. Yet the conventional language used to describe what happened is inadequate and misleading. It is inaccurate, in a literal sense, to say, "The US invaded Iraq" because not everyone joined the invasion. It would be inaccurate in the contrary direction to say, "The US demonstrated against an invasion of Iraq" because not everyone in the US demonstrated—but a much larger number demonstrated against the invasion than were involved in the invasion. (I'm setting aside the consideration that most US government officials did not refer to an invasion at all, but instead talked about liberating Iraq.)

Governments are complex organisational entities. To say they act, speak, bargain or feel is to liken them to individuals who, in contrast, are assumed to be unitary. If a part of a person's body refuses to cooperate, it is seen as dysfunctional, perhaps dangerous, like cancer. Treating a country like an individual invites the assumption that opponents of government policy are similarly dysfunctional, or even dangerous.

When Bush said, "You are either with us or with them [the terrorists]," he played on this analogy of the country with an individual. This "us"—in this instance "us" is "US"—is treated as unitary, when in reality there is no single "us."

If Bush hadn't been able to draw on the linguistic assumption of government-country unity, he would have had to say, "Either you support US government terrorism policy or you oppose it." That's less punchy and less threatening. It's far easier to oppose policy than to oppose "us"!

The use of country names for government actions can be called "statist language": it linguistically attributes the actions of the state—the government and especially leading figures in the government—to the people, to an entire society. It makes it awkward to talk about internal tensions or dissent.[2]

Statist language is a convention: it is the standard way of writing and speaking, especially about international affairs. Any other way can sound strange or cumbersome. It's easier to say, "Iraq invaded Kuwait" than "Iraqi military forces invaded Kuwait."

This convention can mask citizen opposition to government. Saying "China decided" discourages people from realising or remembering that it was only the Chinese government, and probably just a few people at the top, who made a decision, and that the bulk of the population were not involved or consulted and many of them may not have wanted this decision if they had been consulted.

In systems of representative government, government leaders have the endorsement of being elected, but this does not mean their policies reflect the unified desires of the entire population. The freer the society, usually the more that differences of opinion can be articulated.

---

2 This chapter draws on my article "Statist language," *Etc.— A Review of General Semantics,* Vol. 66, No. 4, October 2009, pp. 377–381. For a sophisticated treatment of language and national identity, see Michael Billig, *Banal Nationalism* (London: Sage, 1995), pp. 87–127.

Statist language is one type of what can be called unitary language, in which a group of entities is treated as a whole. Unitary language is appropriate when groups operate under a command system, such as the human body, or a group using consensus decision-making, so everyone agrees. But whenever there is significant conflict or internal disagreement, unitary language can be misleading. The statement "General Motors condemned the strikers," when the strikers are GM workers, offers a different image than "GM management condemned GM workers."

Unitary language often reflects a hierarchical worldview in which rulers or bosses speak on behalf of their subordinates, whether or not there has been any consultation. In the United Nations, when government representatives speak on behalf of their countries this might be reported as "China said" or "Germany said." In 1994, the government of Rwanda held a seat on the UN Security Council. The Rwandan government orchestrated a genocide beginning in April, but tried to hide this from the outside world. When the Rwandan Security Council representative reported falsely that the killings had stopped, conventional statist language might have expressed this as "Rwanda told the Security Council the killings had stopped." But it certainly wasn't the Rwandan people saying this: they were perpetrators, victims or bystanders of the ongoing genocide.

Another feature of statist language is the assignment of people to countries and vice versa. The people living in France are the French, the people living in Guatemala are Guatemalans, and so forth. Conversely, without the

French there is no France. As noted by Michael Billig, "A form of semantic cleansing operates in these terms: there is no gap between the people and its country."[3] There are a few anomalies in the linguistic binding of peoples and countries. For example, there are no United Kingdonians, and for much of the world "Americans" refers to US people, not inhabitants of South and North America. Generally, the grammatical conventions associating people with countries serve to make the division of the world via national boundaries seem natural rather than the result of political and social action.

Sexist language
Statist language has many parallels with sexist language. A few decades ago, it was conventional in English to use "he" to mean "he or she," to use "chairman" to refer to either a man or a woman in the role of chair, and to use "man" to mean "humans." Male pronouns were standard when referring to both sexes.

Feminists challenged what they called sexist language. They said male words made women invisible by making readers visualise men rather than both sexes. Male language made it harder to imagine a woman in a role, especially a traditionally masculine role.

Defenders of the convention argued against change, saying that everyone knew that "he" included both sexes and that "he or she" is clumsy and "they" is ungrammatical. They made fun of critics by pointing to the alleged absurdities involved in removing mention of men from

---

3 Billig, *Banal Nationalism,* p. 78.

language: "woman" would have to be replaced by "womon" and perhaps "person" by "perdaughter."

The conservative defenders of sexist language lost, so much so that many writers, in quoting from text written in the 1960s or earlier, painstakingly notate male pronouns with "[sic]" or replace them with "[he or she]" to highlight their awareness of, and perhaps distaste for, the sexist language in the original.

Examples

Statist language is so common that it easy to produce a host of examples. To provide illustrations, I picked an issue of the *New York Times,* the newspaper most commonly cited as setting a standard for others. I chose an arbitrary issue, 8 January 2009, the first day I was able to purchase a copy during a visit to the United States.

On the front page is a story titled "China losing taste for debt from the U.S."[4] Its lead paragraphs include passages such as "Beijing is starting to keep more of its money at home," "declining Chinese appetite for United States debt," "China has spent" and "Beijing is seeking to pay." Of course it is not literally "China" that is "losing taste for debt," because the article makes no mention of debt preferences among Chinese people, but actually top Chinese economic policy-makers. Only later in the article are there more precise references to "the Chinese government," "Chinese businesses" and "China's leadership."

---

4 Keith Bradsher, "China losing taste for debt from the U.S.," *New York Times,* 8 January 2009, pp. A1, A10.

On page A6 is the story "Ex-prostitutes say South Korea and U.S. enabled sex trade near bases."[5] The reference to "South Korea" and "U.S." must refer to military or political authorities, because the average South Korean plays no role in the sex trade and the average U.S. citizen knows nothing at all about U.S. military bases in South Korea, much less the existence of the sex trade—unless, perhaps, they have read this or a similar article.

This story occasionally uses statist language but for the most part uses more precise references. The first sentence is "South Korea has railed for years against the Japanese government's waffling," which doesn't reveal who in South Korea had railed—the government? activists?—but pinpoints the target of complaint, the Japanese government.

In the second paragraph, the article says "Now, a group of former prostitutes in South Korea have accused some of their country's former leaders of a different kind of abuse: encouraging them to have sex with the American soldiers who protected South Korea from North Korea." Note the precision of "a group of former prostitutes" and "some of their country's former leaders" compared to the reference to "protected South Korea from North Korea," which implicitly groups North Korean citizens with the North Korean government as a threat to South Korea, again a single undifferentiated entity.

---

5 Choe Sang-Hun, "Ex-prostitutes say South Korea and U.S. enabled sex trade near bases," *New York Times,* 8 January 2009, p. A6.

On page A12, one of the several stories on the conflict in Gaza is titled "As Gaza battle goes on, Israel is set to negotiate with Egypt on cease-fire."[6] The title refers of course to the governments of Israel and Egypt. The first sentence begins "Israel said Wednesday ..." This common formulation suggests that "Israel" is a person speaking with a single voice. It disguises the diversity of political opinion within Israel over policies and actions concerning Gaza. Although many readers understand this diversity and treat "Israel said" as "Israeli government spokepeople said," the statist shorthand may discourage thinking of the complexity. For those not familiar with complexities of Israeli politics, "Israel said" reinforces a mental image of discrete entities, Israel, Egypt and Gaza.

Paragraph three begins "Israel suspended its military operations in Gaza for three hours ..." Perhaps the Israeli government or military suspended military operations; most Israelis had no say in this decision, and many members of Israeli peace movements would not like to be implicated in any decision to use military force in the first place.

Paragraph five begins "Hamas fired 22 rockets into Israel ..." How many readers would stop to think that perhaps not every member of Hamas supports firing rockets? Certainly not all of them were involved in the firing itself.

---

6 Steven Erlanger, "As Gaza battle goes on, Israel is set to negotiate with Egypt on cease-fire," *New York Times,* 8 January 2009, p. A12.

Elsewhere in the article there is similar statist language, but more precise language is also used, with references to, for example, "the Israeli Army," "the Israeli government," and "the government spokesman." It is certainly possible to write without statist constructions.

These are just a few examples taken from one issue of the *New York Times*. The same observations could be made using news reports from innumerable sources.

Alternatives
Instead of "We invaded Iraq" or "The US invaded Iraq," what would be a more accurate formulation? One possibility is "the US military invaded Iraq" or "The US government launched an invasion of Iraq." Referring to the military or the government helps to direct attention to those acting, thereby allowing that others, including members of the US population, may not be involved or supportive.

The use of a country's name to refer to the government is quite convenient, and alternatives are cumbersome. The obvious alternative to "US" would be "US government" or perhaps "USG" for short. Those who want to be really precise in their language would say that "US government" is still unacceptable, because not everyone in the government supports actions taken in the name of the government—certainly not the invasion of Iraq.

When talking or writing about government actions, it is straightforward to avoid constructions that conflate the government and the people in a country: just avoid any statements that refer to the country acting as a whole. This means not saying something like "China declared" but instead "a representative of the Chinese government declared" and not saying "India is having talks with Pakistan" but instead perhaps "Indian and Pakistani government officials are having talks." Because the alternatives are cumbersome, it is all too easy to revert to conventional expressions.

Another option is to use the abbreviated form but in an unconventional way. You might say "India opposed the trade agreement" when actually Indian policy-makers supported it—however, only those who are knowledgeable about the issue will understand that you are referring to civil society groups or popular opinion, not the government.

Statist language brings a pervasive bias into reporting, especially on international affairs, typically favouring governments over opponents and popular movements and sometimes over popular opinion. Using different expressions is not easy: habits run deep. Challenging those habits is a small step towards better understanding and better strategic thinking. Non-statist language will not solve the world's problems but it can help make them more apparent.

9
Citizenship

Robert Jovicic was born in France 1966 and came with his Yugoslavian parents to Australia at the age of two. He grew up Australian. His parents became Australian citizens but Jovicic never bothered to do so, because he had permanent migrant status. But it wasn't as permanent as he might have thought. Jovicic became involved in criminal activities. After spending time in jail, he was deported from Australia, to Serbia, where he was unable to work (having been given only a short visa) and didn't know the language.

Jovicic was vulnerable to expulsion from Australia because he lacked citizenship. If his parents had been in Australia when he was born, he could have remained in Australia despite any crimes.[1]

Most people in the world are a citizen of a country; some are citizens of two or more countries. Being a citizen normally means you have the right to reside in a country. Usually you can obtain a passport and travel to other countries.

Citizenship is a key tool used by governments to control populations. If you are not a citizen of any country, you are "stateless" and at risk of being sent somewhere you don't want to go, or even imprisoned.

---

1 After publicity about his desperate plight, Jovicic was able to return to Australia and be granted permanent resident status.

A century or more ago, citizenship was not such a big deal. Relatively few people travelled a lot, but for those who did, there were fewer controls. Passports are a recent invention.

The very idea of citizenship reflects identification with a state, indeed it assumes the existence of states. Without a state, you are a person. When subject to the administration of a state, and accepted as one of its subjects, you are a citizen. As a citizen, you have some rights and privileges not available to those who are not citizens—called aliens. Perhaps it is no coincidence that people from outer space are called aliens. They are not subjects of governments of the planet earth.

The control function of citizenship is most apparent in the plight of refugees. People under threat in their own countries due to war or persecution seek asylum somewhere else, but acceptance is not automatic: they have to be assessed and certified as refugees, and even then they may be kept in camps and prevented from full membership in the receiving country.

Australia illustrates some of the worst practices regarding refugees. Except for Aborigines, the descendants of people who inhabited the continent for tens of thousands of years, nearly everyone in Australia is either a recent immigrant or a descendent of immigrants since the first white settlement in 1788. Despite Australia being a nation of immigrants, recent governments have demonised refugees arriving by boat. They are intercepted by the navy and either pushed back to their port of departure or taken to detention camps in various locations. Those who make it to the Australian continent are also put in camps, sometimes for years, sometimes with no prospect of release. Many of the refugees are escaping conflicts, such as in Afghanistan and Iraq, where the Australian military is involved. The Australian government wants to fight the enemy abroad but not accept responsibility for the human consequences of the conflicts.

Since the early 1990s, Australian governments have demonised asylum seekers in a populist pitch to xenophobic elements of the population. It is a classic case of building in-group support by treating out-groups as dangerous. Although many Australians have relentlessly campaigned against the government's refugee policy, nevertheless both major political parties have continued with the policy, making it ever more punitive, because they believe this wins voter support.

At the same time, the Australian government has run one of the largest planned immigration programmes in the world, on a per capita basis. Hundreds of thousands of immigrants are accepted each year, mainly in two categories: family reunions—existing family members already reside in Australia—and occupational migrants, who bring skills or money to the country. The result of the ongoing immigration programme is that one out of four Australians was born outside the country, from a range of countries: Britain, New Zealand, China, India, Philippines, etc. The parents of many other Australians were born outside the country, most notably as part of the post-world-war-II wave of immigrants coming from Britain, Italy, Greece, Egypt and elsewhere.

So there is a contradiction at the heart of the Australian government's treatment of immigrants. Those coming through formal channels are welcomed; those coming by sea as refugees are portrayed as a threat to the country.[2]

Most people prefer living where they are. They have ties to family and friends, cultural associations, local knowledge and many other connections to their community. Most refugees are fleeing violence, exploitation or extreme disadvantage. Most would prefer to stay in their homeland if it could become stable, safe and prosperous.

The "open borders" movement argues in favour of eliminating barriers to people moving to different parts of the world.[3] To most people, this sounds totally impracticable. Millions of people would immediately want to move to the richest countries. But of course a switch to open borders would not happen overnight. Imagine this scenario. In 20 years, barriers to moving between countries would be removed. There would be intense pressure from rich countries to end the conflicts that generate so many refugees—for example in Afghanistan, Iraq, Sri Lanka and Syria—to challenge repressive rulers and to implement policies to eliminate corruption and enable people to make a decent living through honest labour. Taking these steps would dramatically reduce incentives to move to other countries. They would also reduce internal migration, a serious problem in many countries.

Ending conflicts, promoting responsive government, eliminating corruption and promoting prosperity are

2 Other contradictions in the treatment of immigrants are covered in chapter 11, "Trade deals and tax havens."

3 http://openborders.info; Teresa Hayter, *Open Borders: The Case Against Immigration Controls* (London: Pluto, 2000).

exactly the supposed goals of world development, but there is little pressure on rich countries to push in these directions. Indeed, many major conflicts are either initiated by western governments (think Afghanistan and Iraq, among others) or simply ignored (Congo, Burundi, among others). Rich-country economic policies have served to exploit poor peoples of the world, through a range of measures, while massive corruption undermines prospects for economic improvement.

For the moment, the idea of open borders is a utopian vision that can serve to stimulate thinking and direct action towards a different sort of world, one in which controls over poor people are replaced by controls over exploitative practices. The idea of open borders is also useful when thinking about tactics concerning citizenship that serve the state—or challenge it.

Promoting country loyalty via citizenship
Let me start with the perspective that citizenship can serve state elites by encouraging people to identify with *their* country and state. What methods are used to do this? The first is exposure of citizenship itself. This mainly occurs by a contrast with non-citizens. Probably the majority of people in most countries never even think of citizenship as it applies to themselves: they simply take it for granted. It becomes to their consciousness only when outsiders—immigrants or refugees—seek citizenship. It also enters awareness when travelling to areas where passports and visas are required. In some countries, citizenship must be verified before being able to vote or undertake certain

jobs. In filling out forms, you may have to indicate your citizenship.

The second promotion tactic is valuing. Many people may take their citizenship for granted or treat it in a purely pragmatic manner, as a necessity for getting around, something like packing suitable clothes for a trip or obtaining a trade qualification in order to get a job. However, for others, citizenship is a matter of great pride. Governments foster this for new citizens, in special ceremonies. More generally, patriotism is commonly intertwined with valuing citizenship, as a symbol of a connection legitimised by government. Furthermore, many people may come to think of citizenship as an achievement or highly desirable attribute, as something special about themselves, rather than as an arbitrary designation that is created and administered by governments.

The third promotion tactic is explanation or, in other words, giving reasons for citizenship. Among legal scholars, the rationales for citizenship are discussed, but for the general public, there is little discussion of citizenship as a system. Instead, most commentary is about who gets to be a citizen, who is excluded and the justifications for different treatment. For example, the Australian government justifies its immigration programme mainly in terms of the national interest, with two main groups: business immigrants, who bring cash and business skills, and family reunions. There are various debates about these, and complaints about abuses of the system, but seldom any questioning of citizenship as a system of controlling movement.

The fourth promotion tactic is endorsement. Governments give their official support to citizenship arrangements, with various formal processes associated with them: employment restrictions on non-citizens, issuing of passports, citizenship ceremonies, and the various patriotic events and rhetoric. Citizenship is a key means of demarcating an in-group, citizens of a country, from an out-group, everyone else.

The fifth promotion tactic is rewards. Being a citizen has quite a few advantages, depending on the country, for example being able to come and go, have jobs, receive welfare benefits and undertake lower cost education. Most people born in a country and who remain in it take these advantages for granted, but for others, gaining the benefits of citizenship is a major issue, especially for those without a lot of money, education and connections.

In summary, citizenship is one of the elements of the complex of practices and ideas that cement many people's identification with a country. This means in practice association with the country's government, because the government sets and administers the rules for citizenship, in accordance (usually) with international agreements between governments. Citizenship serves to control people's movements in a world where travel is easier than ever before and where restraints on the movement of capital have been dramatically reduced.

Citizenship thus is caught in the middle of some deep contradictions. Governments are committed to the system of citizenship because it gives them power, but it also is a potent trigger for suspicion and even anger at out-groups, including non-citizens who engage in commerce, for example buying property or selling goods in competition with locals.

Alternatives to citizenship

Quite a few people don't really care about citizenship. If you were born a citizen and never travel anywhere requiring a passport, being a citizen may not seem important. Others treat citizenship as a pragmatic matter, something necessary to get a job and move around, and have no particular attachment to the country or countries of which they are citizens.

Then there are a few people who envisage something different. They might prefer to think of themselves as a citizen of the world, a "global citizen," with primary loyalty to all humans, or perhaps the biosphere or the planet, including everything from air to rocks. The implications of an alternative model of citizenship can be a matter for discussion. Does this mean freedom to move to any part of the world? Or could a person only settle in an area if invited by local residents? What about services now provided by governments, such as unemployment payments? Does global citizenship imply dissolution of governments, or only that governments have to adapt to free movement of citizens?

One possibility would be to look at the arrangements within the European Union, which allow free movement, without passport controls. The New Zealand and Australian governments have removed restrictions on movement between the two countries. Could such arrangements be gradually expanded to more parts of the world?

For the purposes here of looking at tactics, it can be useful to look at a particular alternative. However, alternatives to citizenship are so far off the mainstream agenda that it is not necessary to specify details. There is virtually no public discussion of alternatives to the conventional model of citizenship. For those with money and skills, there is considerable mobility, and citizenship is not a rigid restraint. For those fleeing wars, exploitation, discrimination or poverty, the citizenship system is a barrier to finding a safe haven. It is for this latter group that public discussion of alternatives is hardly ever discussed as a serious option. So the first tactic against alternatives is a de facto cover-up.

Next is denigration of alternatives. To the extent that the idea of open borders is even acknowledged, it is usually dismissed as unrealistic if not dangerous. More revealing, though, is attitudes towards those seeking to move to other countries but not welcome by governments. Legitimately, they can be called asylum seekers or refugees, or migrants seeking a better life. They are also given derogatory labels. In Australia, people who attempt to arrive by boat seeking asylum are commonly called illegals, even though what they are doing is legal according to international law. They are called queue-jumpers, even though there is no queue for seeking asylum. They are called economic migrants (often with a contemptuous tone of voice), suggesting they have no justification to migrate, even though other sorts of economic migrants, who have more education, money and connections, are welcome. Sometimes, it is even suggested that asylum seekers are criminals or terrorists.

Because alternatives are not on the agenda, there is not much public discussion of them. If open borders entered the public debate, then undoubtedly arguments would be raised against the possibility, but for the moment the discussions remain among academics. Similarly, there seems to be little need to take action to dampen enthusiasm for open borders through formal investigations or intimidation of proponents. In Australia, the dominant discourse is driven by policies on refugees. Opponents of the government's policies typically argue in terms of international agreements concerning human rights, not in terms of alternatives to citizenship.

Challenging the citizenship-patriotism connection
Because citizenship is so often taken for granted, a first step in challenging usual assumptions is to point out contradictions in the uses of citizenship, for example the different way the rich and poor are treated.

One of the key flash points in citizenship struggles involves responses to immigrants: people seeking to change their residence and sometimes their citizenship. In quite a few countries, governments put tight constraints on acceptance of "unwelcome" immigrants. Pushing for fair treatment of asylum seekers is an attempt to ensure that international agreements are followed. There are many campaigners involved in supporting the rights of refugees.

However, there is another side to the issue: governments pushing for free movement of capital and the selection movement of labour to serve corporate agendas. Highly skilled or wealthy individuals receive a welcome seldom extended to asylum seekers arriving outside the usual protocols. Questioning the free flow of capital can buy into a nationalist agenda. It is not so obvious whether or how this challenges the systems of citizenship and patriotism.

Rethink
It seems like there are two categories of citizenship, or perhaps two categories of citizens. People who have plenty of money and connections experience no barriers to travel and to being able to live in other countries for short or longer times. These are people who have the mobility of capital: barriers have been removed, so they have various options for deploying their labour. Call this category P, for privileged or professional.

People in the second category have insufficient money, skills or connections to move to more desirable parts of the world. This category includes refugees. It also includes people who are tied to land (farmers), to family networks (through obligations) and to local sets of institutions. People in the second category have limited mobility; the cost in trying to move can be enormous, both financial and associational. Call this category R, for residential or restricted.

There seem to be different ways of thinking about these two categories of people. P-people are welcome, at least in some places, whereas R-people are unwelcome except in special circumstances. Governments typically welcome P-people but create barriers to R-people.

For P-people, citizenship becomes a secondary matter, because it does little to restrict movement or work. For R-people, citizenship is a crucial form of control.

Nearly all the scare-mongering about immigration and refugees is about mobilising concern by local R-people against R-people from elsewhere.

Double standards
The use of citizenship as a method of control contains an intrinsic double standard. First is the standard applied to those without money, skills and connections. They are citizens of their own country, but have little prospect of gaining citizenship in another country, except through enormous efforts and sometimes extreme sacrifice.

For many governments, these sorts of people are undesired as potential immigrants. Furthermore, many citizens identify with their governments and see the poor people of the world as undesirable intruders, who should stay where they are. This fear of foreigners is often linked to racism. It has become almost an inevitable accompaniment to nationalism and country-centreness. Politicians can promote this sort of xenophobia as a means of building support, and because of the level of popular support for measures against these sorts of immigrants, some politicians fear to move too far in other directions.

However, there is another group of people: those with money, skills and connections. For many practical purposes, they are free to move to other countries for visits, jobs and permanent residence. Though how easily they can do this depends on the person and the country, billionaires usually have more options than millionaires.

Conclusion
Citizenship is a crucial element of the way the world is divided into countries, each administered by a government. If you're a citizen, you're part of a recognised unit—a country. If you're not a citizen, you're called "stateless" and are much more vulnerable to ill treatment. Hence there is a great incentive to be or become a citizen, thus reinforcing everyday nationalism and the governments that benefit from it.

10
Our economy

In Australia, the government and the media give extraordinary attention to the state of the economy. One of the most common talking points is jobs. "The jobless rate has increased from 4.8% to 5.0%. The government needs to take action." "Two hundred thousand new jobs were created in the past three months."

Loss of specific jobs can be a source of alarm. "We need to provide support [meaning a government subsidy] to the car industry, otherwise hundreds of jobs will be lost." "A factory just closed, and 25 workers lost their jobs."

There are lots of things to question concerning the jobs mantra. Seldom does the government talk about opportunity costs: a tariff or government subsidy to manufacturing industry could instead have been provided to a different sector, perhaps saving more jobs. Massive investment in mining of iron ore or uranium might have created several times as many jobs if put instead into solar power and energy efficiency.

One of the assumptions in these discussions is that the goal is more jobs. Hardly ever is there discussion of whether these jobs are satisfying, secure or full-time—all very relevant considering that many new jobs in Australia are part-time and not permanent. Government statistics are based on the arbitrary definition that if you work at least one hour per week, you are "employed." This minimal requirement is designed to boost the figure for the employment rate.

An alternative perspective is that the goal should not be jobs, but rather satisfying work for everyone, and furthermore that paid work should not be the primary way in which the allocation of the economic product—who gets what—is determined; instead, allocation should be according to need. This radical view is hardly ever articulated by mainstream commentators.

Set all this aside, and consider one additional assumption underlying commentary on jobs: the focus is always on Australian jobs. Never does anyone talk about the value of creating more jobs in other countries, especially in poor countries. Discussions about "the economy" are about the Australian economy, or more generally about the world economy and the economies of Australia's largest trading partners. Almost completely absent is anyone asking, "How can Australian economic policy help poor people of the world?"

The government focuses on boosting the Australian economy, or sometimes boosting the fortunes some certain groups within Australia, most commonly those better off. Investors, such as investment funds, focus on returns for themselves or their members. Trade unions focus on jobs, wages and conditions for their members.

Concern about world poverty is commonly seen as separate from concern about the Australian economy. World poverty is treated as a matter for the foreign aid budget or for voluntary organisations like World Vision or simply for someone else, such as the United Nations, or perhaps for the governments of other countries. Alternatively, world poverty is seen as something that will be addressed by improvements in the world economy, in the usual trickle-down effect: as the rich become richer, some of their wealth will provide jobs and greater income to the poor. Meanwhile, though, attention is continually directed to what is good for Australians.

Occasionally, there is some thought to workers elsewhere—they are stealing Australian jobs! When call centres are closed in Australia and the work taken to India, there are lots of gripes about loss of jobs and poorer service, with only occasional mention of the benefits for workers elsewhere. Similarly, Australian exports are seen as a good thing because they bring money into Australia, and only secondarily because they are providing a service or product to others.

Overall, discussions of the economy within Australia are incredibly country-centred: they are almost entirely about what serves Australians. The government is seen as the key player in the economic arena, designing policies that will serve Australian businesses and workers.

Sometimes attention is drawn to regional or local jobs and economic performance, such as job loss in the state of New South Wales or the city of Sydney. There might be reports about economic growth in China: because there is so much trade between Australia and China, the Chinese economy affects the much smaller Australian economy.

In Europe, discussions of "the economy" might be either about a country such as Italy or Germany, or the EU more generally. In the US, the state or local economy will

receive some attention along with the US economy generally. In neither case is there much awareness of how things
are going in Peru, Cambodia or Zambia—they are off the
radar.

The relative attention to jobs and economic performance can be seen as a form of competition for attention.
Those with the most power and influence try to make
people aware of things from their perspective. Governments seem to have the greatest influence, with mass
media usually following government priorities. This
process can be looked at in terms of tactics.

1. Exposure. Governments collect and publicise
statistics about the national economy, and to a lesser
extent local economies.
2. Valuing. More jobs and greater economic growth
are always seen as a good thing, while other priorities, such as happiness, equality or the environment,
are secondary.
3. Explanations. Attention is focused on paid work
and economic indicators are treated as signs of what
is most important.
4. Endorsement. Governments and various agencies
make authoritative announcements about jobs and the
economy, giving this emphasis a stamp of approval.
5. Rewards. Those who go along with the dominant
framing—especially those who collect and interpret
statistics and who write about the economy—can
expect a receptive and sympathetic audience.

Australian news reports often tell whether the share market has gone up or down, and give the latest exchange rate with the US dollar. Sometimes reports tell about economic trends in major countries, especially China and the US. The economies of poor countries are almost never mentioned.

Another aspect of self-centredness in public discourse about the economy is the issue of "Australian-made." Decades ago, it was a matter of pride for some Australians to buy a Holden, the General Motors car manufactured in Australia. Buying a Toyota or some other car from Japan or Korea was somehow not supporting Australia. Those days are long gone: foreign cars are common, and most people buy the cars they think are the best value, which means the Australian car-manufacturing industry is collapsing.

Nevertheless, there are complaints from some sectors when the government is said not to be doing enough for Australian businesses. When tariffs were reduced on some products, such as clothes and food, imports boomed. Yet, because of residual loyalty to the idea of being Australian, some companies advertise themselves as being "Australian owned" or "proud to be 100% Australian," though some of these claims are dubious.

Then there are concerns raised when foreigners buy properties and businesses in Australia. This is sometimes presented as a foreign threat. There is a bit of racism involved: there may be concerns raised about Chinese investors buying Australian land, but none about British investment. There is a clear double standard too, because some Australian-based multinationals have bought properties or companies in other countries, with never a peep of protest from commentators, except perhaps that it might be unprofitable.

There are two competing pressures on discourse about imports and exports. One is to maintain Australian ownership and to buy Australian-made goods; the other is to buy whatever is cheaper or better, whether made in Australia or elsewhere. The thrust of capitalist markets is towards greater international competition, so the appeal of being "Australian-made" has been declining. The key thing to note is that both discourses put Australians at the centre, as beneficiaries either as workers and owners of Australian businesses or as consumers of services and products. There is little thought in either configuration of thinking about benefiting people in other countries, except as a spin-off of world economic growth.

A question arises: how does thinking about "the economy" become so centred on benefits to the home country? The primary driver is the national government, where decisions are made about government expenditure, trade agreements, investment guidelines and the like. It is in the government's own interest to build the national economy: after all, the government obtains revenue by extracting it from the economy. Regional and local governments want to develop their own economies, but they have fewer resources to do so.

The Australian economy is semi-closed. Capital can move fairly freely, but labour cannot. People in Australia can move to different parts of the country in search of better jobs, among other things, but immigration to Australia is restricted. It is a central contradiction of global capitalism, with its rhetoric about free markets, that people cannot readily move across borders in search of jobs. The result is an orientation to the economic entity in which people can move. In Europe, this is the EU, so there is a division of attention between national economies and the EU economy. The Australian government has little incentive to think more broadly in terms of its goals.

Economics in the media: an example
My comments here about the self-centredness of Australian economic discourse are based on observations over many years. To back up these generalisations, it would be necessary to carry out interviews, analyse media discourse or listen to focus group discussions. Here, more modestly, I only *illustrate* the Australian media's orientation to the Australian economy. I picked an issue of the *Sydney Morning Herald,* considered one of Australia's quality newspapers, choosing the issue of 23 February 2015, the day I wrote the first draft of the preceding text. My comments here are about articles concerning economics in the news pages.[1]

On page 2, there was an article about whether workers in pubs should continue to be paid higher wages on holidays declared by Australian states, in addition to the standard eight national public holidays. For example, the government of the state of Victoria declared a public holiday on the weekend of the grand final of the Australian Rules football competition. The Australian Hotels Association, representing pubs and clubs—where food and

---

1 A separate analysis could be undertaken of the business pages.

alcohol are served—wanted uniformity across the country in what are called penalty rates of pay. This was a story about national economic matters.

On page 3 was an article about a scandal in the National Australia Bank, one of Australia's four major banks. NAB financial advisers had been exposed for offering misleading advice to customers, causing the customers to lose large amounts of money. Due to additional leaked documents about the scandal, adding to previous exposés, there were calls for a royal commission into Australia's financial planning sector. This was a story about national economic matters.

A brief report on page 6 began "Access to affordable medicines could be under threat in Australia if the US gets its way in secretive negotiations over a trade deal involving 12 Pacific-region countries, academics have warned." The theme was risks to Australian patients. The report concluded with the statement "Trade Minister Andrew Robb said he would not agree to anything that was against Australia's interests." Both sides in the argument about the trade agreement thus used the rhetoric of benefiting Australians.

An article titled "Annual coal health toll $600m, doctors say" reported on estimates of damage to health in the Hunter region, around the city of Newcastle. The article highlighted a clash between economic benefits to the state versus health costs. "A 2014 report for the NSW Minerals Council estimated Hunter coal contributes in total $6.3 billion annually to the state's economy, or almost half of the total mining industry's output in the state. The region's coal industry also employed more than

18,000 people." The orientation is to the economy of the state.

On the comment pages, the editorial for the day addressed the issue of the federal government's payments for childcare assistance, saying "Taxpayers subsidise childcare by almost $7 billion a year," something that "helped the economy by allowing more mothers and fathers to balance work and parenting, which in turn has increased productivity, economic growth and living standards." However, the editorial stated, this system may not be efficient. The assumption is that childcare policy should be about benefiting the Australian economy.

Among the letters to the editor, a section was devoted to ones about housing. An article the previous day, titled "Rich pensioners may be too much at home," raised concerns about people owning million-dollars homes and receiving the aged pension: perhaps they should move out. Letter-writers contested this, for example pointing out that the median house price in Sydney was approaching a million dollars. The assumption underlying this debate about the economics of retirement was that the trade-off between what was fair to individuals, in particular elderly homeowners, and fair to the Australian taxpayer.

Among the letters, there was one offering a contrast to the usual emphasis on money: Jenny Blake commented that, "… the joy of being part of your grandchildren's lives can never be calculated in dollars and cents. It is a sad society we have become when everything is measured by money."

Challenging economic self-centredness
There are quite a few ways to challenge the orientation of economic thinking and discourse towards self-interest, with "self" often involving an identification with the country and the state. These can be generally classified into methods of confrontation, fostering alternative identifications, and putting priority on different goals.

Methods of confrontation directly challenge the standard orientation of economic thinking. The Occupy movement's slogan of "We are the 99%" is an example: it switches the orientation from economic growth to economic inequality. Then there are those who raise awareness about poverty and disadvantage. They expose crimes of the wealthy, point to exploitation of workers, oppose trade deals that benefit the rich, and question the world system of trade and debt. There are actually lots of people pushing for a different set of priorities and who provide a different agenda than the usual one built around the rhetoric of economic nationalism.

A second set of methods seeks to promote identification with a different group than the country, region or city that is the usual focus of economic discourse. The traditional socialist emphasis on the working class is a classic example: the working class transcends national boundaries and pits workers against the ruling class, thus questioning economic nationalism with a different focus. However, working-class consciousness often is linked to wages and conditions of workers, and thus feeds into the preoccupation with what is good for workers—in this country. Trade unions officials usually put the interests of unionists in their union first, above other considerations. Seldom do they make decisions with a primary concern about workers worldwide. Those who are unemployed or in non-unionised sectors of the world economy are not often of great concern.

Rather than identifying with workers, another possibility is identifying with poor people worldwide—even if you are not one of them. This is the approach of those concerned with poverty reduction, movements against exploitation, campaigners for rights of the most disadvantaged, public health advocates, and various others. Whether identifying with poor people is an effective counter to economic nationalism is probably best assessed on a case-by-case basis.

A third set of methods to challenge state-centred economic thinking questions the assumptions in conventional growth economics. An example is the idea of a steady-state economy, namely one that doesn't grow any more. The steady-state economy is a long-term necessity, at least when growth involves tangible things like energy and consumer goods, simply because eventually resources and non-renewable energy sources will be exhausted. Therefore, it makes sense to start planning for a steady-state now.[2]

Research on happiness offers another way of questioning normal thinking about economics. Greater income does, on average, lead to higher reported happiness, but only up to a point. Above a modest standard of living, greater incomes lead to little or no

2 A classic reference: Herman E. Daly (ed.), *Toward a Steady-state Economy* (San Francisco: W. H. Freeman, 1973).

increases in reported happiness levels. One striking finding is that in countries like Britain, Japan and the US, recorded average happiness levels have hardly changed nationwide over several decades, while the per capita gross national product has greatly increased. What this means is that people are earning more and they have bigger houses, nicer cars and more electronic gadgets— but they are no happier, on average, than earlier genera-tions who were, by today's standards, deprived.[3]

The interpretation of these findings has been con-tested, but what all economists accept is that money has a declining marginal utility: an extra dollar means a lot more to a poor person than to a billionaire. The implication, in terms of collective welfare, is that there are greater benefits from bringing people out of poverty than in adding to the wealth of those already well off. In other

---

3 Gregg Easterbrook, *The Progress Paradox: How Life Gets Better While People Feel Worse* (New York: Random House, 2003); Bruno S. Frey and Alois Stutzer, *Happiness and Economics: How the Economy and Institutions Affect Wellbeing* (Princeton, NJ: Princeton University Press, 2002); Bruno S. Frey in collaboration with Alois Stutzer, Matthias Benz, Stephan Meier, Simon Luechinger and Christine Benesch, *Happiness: A Revolution in Economics* (Cambridge, MA: MIT Press, 2008); Richard Layard, *Happiness: Lessons from a New Science* (London: Penguin, 2005). There are some who dispute this finding. See, for example, Michael R. Hagerty and Ruut Veenhoven, "Wealth and happiness revisited—growing national income *does* go with greater happiness," *Social Indicators Research,* 64, 2003, 1–27, and subsequent articles by Richard Easterlin and by Veenhoven and Hagerty.

words, a more equal distribution of income and wealth should be the goal, rather than increases in gross domestic product: growth (progress) in equality, not growth in wealth. Research suggests this will increase overall happiness. Indeed, people who are materialistic, seeking ever more income and possessions, tend to be less happy than average; therefore, fostering a more caring and less acquisitive society would improve wellbeing overall.[4]

Then there are particular activities that usually increase personal happiness, including helping others, expressing gratitude and being physically active. These do not require much money, and just about anyone can undertake them. Potentially, they provide an alternative direction for economic priorities.

Much of the research on happiness—also called flourishing or wellbeing—is oriented to the individual, which has the disadvantage of meshing with individualism in materialistic striving. However, it is possible to rethink some of the happiness-promoting activities as collective endeavours, and furthermore ones that lead to social changes. For example, helping others is a potent method of improving one's own happiness, as long as this doesn't become routine or obligatory. Designing a society around enabling people to help each other directly—without government as the intermediary body, collecting taxes and providing welfare services—offers more prospects for happiness.

---

4 Tim Kasser, *The High Price of Materialism* (Cambridge, MA: MIT Press, 2002).

This leads into economic alternatives, of which there are many. Some alternatives involve a greater role for government, other less. For example, a guaranteed annual income is usually assumed to be provided by government, whereas local currencies reduce the role of the central government.

One of the most promising economic alternatives is building the commons, namely the resources that are freely available to everyone. Two traditional types of commons are libraries and public parks. Anyone can check out a book from a library or, these days, use the Internet. Anyone can visit a park area in a city. The history of libraries and parks is instructive: workers had to struggle to introduce and maintain these facilities.[5] After all, they are competitors to private enterprise. It is possible to imagine a world without libraries, but instead only bookshops and Internet cafés, and a world without public parks, but instead only privately run parks charging substantial fees for entrance.

With the development of computing and the Internet, a new type of commons has emerged, referred to as the digital commons. Its best-known feature is free software, such as the operating system Linux. Free software is produced by collectives or networks of programmers who provide their services without charge, and the resulting products are available to anyone. One of the slogans of the free software movement is "free as in free speech, not free beer." The key to free software, and its close relation open

5 Raymond Williams, *The Long Revolution* (Harmondsworth: Penguin, 1965), pp. 73–74.

source software, is that the code is publicly available, so anyone can use it or modify it, but not copyright it.[6]

The open source movement has inspired parallel developments in a range of areas. For example, there are now open-source colas, with the ingredients displayed on beverage containers, and open-source code to run 3D printing, an alternative to regular manufacturing.

The open source movement is expanding the role of the commons, and is thereby providing an alternative to government as a source of economic welfare. The commons is a more general alternative to the economic role of government, which is to collect taxes and provide both individual and collective services. Governments can support commons, as in the case of libraries and public parks, but in other cases governments oppose commons and instead support corporations and their efforts to undermine or outlaw commons. This is apparent in government support for expansion of intellectual property regimes that protect the monopoly-privilege positions of software companies, pharmaceutical manufacturers, large book and journal publishers, and Hollywood producers.

---

6 Samir Chopra and Scott D. Dexter, *Decoding Liberation: The Promise of Free and Open Source Software* (New York: Routledge, 2008); Karl Fogel, *Producing Open Source Software: How to Run a Successful Free Software Project* (Karl Fogel, 2005); Glyn Moody, *Rebel Code: Linux and the Open Source Revolution* (New York: Basic Books, 2001); Steven Weber, *The Success of Open Source* (Cambridge, MA: Harvard University Press, 2004). The differences between free and open source software and their associated movements are important but are not central to the discussion here.

In summary, there are at least three approaches to challenging economic nationalism: confronting economic self-centredness by questioning standard assumptions and silences, for example as done by the Occupy movement; promoting identification with a different group, such as local government or the working class; and questioning assumptions underlying conventional thinking about economics, as in research on happiness and in the commons as the basis for an economic alternative. All of these are occurring and, to counter them, governments remain active in shaping discourse.

11
Trade deals and tax havens

Globalisation is the process by which goods, services and all sorts of activities become spread about the world rather than restricted to particular localities. Globalisation can occur with all sorts of things. Stamp collecting is a global activity, and many collectors trade stamps with people in other parts of the world.

The controversies over climate change, nuclear power, fluoridation, vaccination and pesticides are globalised in the sense that the same sorts of arguments, participants and actions are found in different places, and there is considerable sharing of information and ideas between groups in different places.[1] There are differences, to be sure. For example, in countries with nuclear power stations, opponents focus more attention on reactor accidents and long-lived radioactive waste than in countries with no nuclear facilities. In some places, where nuclear power has never been a serious option, there is little debate about it. Globalisation does not mean that exactly the same ideas, activities or products are found everywhere, but rather there is a process by which similar developments occur in many places, often with adaptations to the local circumstances.

---

1 Brian Martin, "The globalization of scientific controversy," *Globalization*, Vol. 7, No. 1, 2008, http://globalization.icaap.org/content/v7.1/Martin.html

Sport is increasingly globalised. There is one major global sport with a huge following, football (otherwise known as soccer), and many others for which there are international competitions, such as table tennis and swimming.

The English language is gradually becoming a global language, becoming the dominant second language in many countries in addition to countries where it is the first language.

Globalisation is not always a good thing. Organised crime is increasingly global, with ties between syndicates in different countries. In the illegal drug trade, production, distribution and sales often occur across country borders.

Despite globalisation, the majority of most activities in the world occur locally and nationally. Most families, for example, live together rather than being spread across several countries. Most commuting is local. Despite the increasing ease of international travel, many more trips are to nearby locations. Globalisation needs to be seen in conjunction with the contrary process of localisation.

My interest here is in one particular type of globalisation: the rise of large corporations whose operations span several countries, and the associated distribution of goods and services in markets in these countries. This can be called corporate globalisation, to distinguish it from other types of globalisation. Global trade has existed for centuries; corporate globalisation involves an increase in the number and power of corporations, with headquarters in one country, that have significant operations elsewhere. A well known example is McDonald's, whose outlets are found in dozens of countries and whose trademark arch is one of the world's most recognised logos.

Tactics and corporate globalisation
The state and globalisation seem, at least on the surface, to be in tension with each other. The state exercises its power from control over politics and economics within a country, whereas multinational corporations have as primary goals expansion and profits regardless of where they are based. State elites presumably have most to gain by putting state interests first, whereas multinational corporate elites care less about any particular state and more about corporate interests.

This tension is resolved by noting a common interest between elites, namely those with most power and money who are at the apex of political and economic systems. Governments derive much of their income, through taxation, from economic activity that is controlled and stimulated by large corporations. Governments cannot easily act against the interests of the largest corporations, and few politicians want to, because they are lobbied by corporate representatives and usually subscribe to a capitalist ideology. Similarly, corporations depend on governments to provide the legitimacy and coercion necessary to preserve private property and to establish and enforce rules for markets. Without governments, corporations could be challenged by their own workers, undermined by unscrupulous rivals, and lose access to markets.[2]

---

2 Robert L. Heilbroner, *The Nature and Logic of Capitalism* (New York: Norton, 1985), p. 105: "remove the state and the

A convenient way to understand the relationship between states and multinational corporations is in terms of a cooperative alliance of those with the most power and wealth against those with much less. However, there are tensions in the relationship, with leaders of the state and of multinational corporations being pulled in different directions by the logics of their respective enterprises. One important tension arises from a central contradiction in global capitalism, concerning the mobility of capital and labour. By the logic of capitalism, both capital (money for investments) and labour-power (workers) should be able to move freely, so that capital can be deployed in areas of greatest profitability and labour can similarly move to where wages are highest. According to neoclassical economic theory, this increases overall productivity to the greatest extent. Accordingly, leaders of multinational corporations have pushed against any restrictions on where and how they can run their operations, and finance capitalists have sought freedom to move money about as they wish. In this context, the so-called Tobin tax, a tiny percentage tax on any cross-border flow of money, is a radical proposal, because it would put a brake on the most volatile forms of financial speculation.

What then about the mobility of labour? Companies usually prefer to have access to labour at lower wages if skills are similar. One option is immigration; another is guest workers, who come from lower-wage countries but

regime of capital would not last a day." See also Michael Moran and Maurice Wright (eds.), *The Market and the State: Studies in Interdependence* (London: Macmillan, 1991).

do not gain citizenship. For example, most of the routine work in Saudi Arabia is done by millions of guest workers from India, Pakistan, Egypt, the Philippines and other countries. However, state elites put severe limits on the movement of people who otherwise like to serve the desires of corporate leaders. The reason is that state loyalty is served by fostering a sense of us versus them: the in-group bias of humans can be harnessed to build loyalty to the abstract entity of a country.

Think of it this way. If states did not exist, and there were no border controls or requirements for passports, then people would be free to move wherever in the world they liked, with the primary constraint that there was a place for them in a new location. Most people would probably prefer to stay near those among whom they grew up and built relationships, but some—especially in areas of exploitation and violent conflict—would prefer to move.

This does not work when there are states that create their own rationale by providing services to a population, such as education and military defence, while maintaining various forms of control over the population in order to extract a surplus (through taxes and other means). Unrestrained movement of people disturbs the connection. If people can move freely, they are less likely to be susceptible to the methods that state supporters use to build identification with a country and its government. If they travel widely, they are exposed to a variety of political leaders and systems and may decide that the one they grew up with could be improved.

Those with money and desirable skills have, for centuries, been better able to move across boundaries, and some of them have developed global perspectives as a result. But until the last century, most long-distance travel was slow and expensive, and hence restricted. There has been extensive migration, for example from Europe to various colonised parts of the world, and from Africa to life elsewhere as slaves. Mobility is nothing new, but the ease of going back and forth within weeks or days is unprecedented.

Cheap and easy mobility poses challenges to state administrators. The extensive use of identity cards (passports) is a recent innovation, introduced by states seeking to establish themselves as the only legitimate controller of people's movements.[3]

All this suggests that the contradiction between the mobility of labour, which would serve capitalists, and control over the mobility of labour, which serves states, has become ever more acute. This contradiction reveals itself in the different methods used by governments in relation to trade agreements.

Trade deals
Global trade has existed for millennia, well before the rise of the state system and the imposition of border controls. The industrial revolution and the emergence of modern

---

3 John Torpey, *The Invention of the Passport: Surveillance, Citizenship and the State* (Cambridge: Cambridge University Press, 2000).

states happened over the same period, and each transformation shaped the other.[4]

The usual thinking about trade is that it is mutually beneficial. Why then would a government set up barriers to trade, such as high tariffs or prohibitions against imports of specific goods? The reason is to protect local enterprises from foreign competition. Unrestricted trade, combined with protection of private property, typically results in the development of oligopolies and monopolies. Without restrictions, these could spread across boundaries, engulfing local businesses. Protectionism enables a local economy, under government or local business ownership, to survive and expand until ready to compete internationally. As a rule, free trade serves those with the greatest economic power.

So we come to contemporary trade agreements. They are often called *free* trade agreements, but this is misleading because they usually contain various restraints on trade, including quotas and intellectual property protection, and none enable significant mobility of labour. The label "free trade" is useful to proponents because it suggests that everyone will benefit while disguising the mechanisms that restrain local decision-making. For example, the North American Free Trade Agreement (NAFTA) contains provisions allowing corporations to sue governments over any law or regulation that hurts profits. Many of the legal actions initiated under NAFTA's Chapter 11 are against the Canadian govern-

---

4 Charles Tilly, *Coercion, Capital, and European States, AD990-1992* (Cambridge, MA: Blackwell, 1992).

ment due to its environmental regulations. Critics have said that trade deals enable corporations to override the policies of sovereign states.

Many trade deals mainly benefit the powerful groups in the stronger parties to the deals. In the US, Congress extends the duration of copyright whenever it is about to expire, so for books, it is now 70 years after the death of the author, with a related term for corporate works. This extension of copyright has been dubbed the Mickey Mouse Protection Act, because it retains the Disney Corporation's intellectual property rights over the cartoon figure of Mickey Mouse, which would otherwise expire.

Intellectual property includes copyright, patents, trademarks and trade secrets, among other forms of law that restrict people's use of ideas and their expression. It is a restraint on trade: a copyrighted text cannot be used by others for commercial purposes. The rationale for copyright is to allow a creator exclusive rights for a period of time in order to stimulate creative production. The duration of copyright, initially quite short, has been extended far beyond any rational basis. Will authors really want to write more novels because they know their heirs (or their publishers) will be able to restrict others from publishing them for decades after their death? What difference will 70 rather than 50 years of post-death protection make to their productivity or creativity? In nearly every case, the benefits from such extended protection flow not to the creator but to non-creators whose control is guaranteed by the government. Such examples make it obvious that intellectual property regimes are in the service of powerful groups, especially pharmaceutical companies, major publishing houses, software companies, and Hollywood producers.

Companies based in the United States gain most of the benefit from these restraints on trade, sometimes called monopoly privilege. In nearly every other country in the world, greater monopoly privilege is harmful as assessed on a national basis. For example, people and local companies in Australia pay vastly more for access to products covered by copyright, patents and other forms of intellectual property than any returns from Australian ownership. In practice, this means that Australians (individuals, companies and the government) pay far more for access to pharmaceutical drugs, proprietary software, books and Hollywood films produced in the US or other countries than returns from its own products similarly covered. However, this did not prevent the Australian government agreeing to extend its own copyright term from 50 to 70 years post-death, something overwhelmingly advantaging US owners over Australian ones.

This is just one example of many showing that trade deals serve multinational corporations over local interests, and that governments will make agreements that hurt national interests. They do it because their loyalties are more to wealthy and powerful groups: they see the world from the perspective of these groups and sincerely believe that their actions will also serve the general interest.

This does pose a difficulty for governments. They need to sell the deals to their own people. They are caught in a dilemma: how to serve the interests of corporate (and government) elites while convincing citizens that they are serving national interests. Few of them think of this challenge in these terms, because they believe they are serving national interests, but in practical terms they have to negotiate the two prongs of the dilemma. To highlight the dilemma, it is useful to look at the usual techniques used by those who take actions that others might see as unfair: cover-up, devaluation, reinterpretation, official channels and intimidation. My expectation is that when perpetrators in these circumstances are trying to serve two contradictory goals, their use of these methods will be inconsistent.

Let's begin with cover-up and its obverse, exposure. In negotiating trade deals, governments collectively operate with great secrecy, not revealing the proposed terms of the agreements. Yet at the same time they trumpet the great advantages of the deals for their citizens. The secrecy—the cover-up—of the provisions of the deals is to hide their damaging aspects from their own citizens, who might be able to mobilise to resist them. (Politicians say secrecy is needed so negotiators can discuss sensitive matters confidentially.) On the other hand, political leaders are quite happy to say how wonderful the deals will be for everyone. The tension between these two stances is bridged by "trust us."

In 1997, while the Multilateral Agreement on Investment (MAI) was being negotiated in secret, a US citizens' group, Global Tradewatch, obtained the text, which had hitherto been kept secret, and circulated it to campaigners in several countries. This exposure was instrumental in the popular efforts to stop the MAI.

You might think that if deals were really so good for everyone, politicians would be pleased to tell everyone about what conditions were being discussed and what trade-offs were being considered. The reality is that many of the points being discussed are unwelcome to citizens, especially to specific groups. When the Australian government was negotiating a "free trade" agreement with the US government, it does not look good to say, "We're going to agree that no Australian-produced sugar will be allowed to be sold in the US." It looks like a restraint on trade rather than free trade.

In 2015, while the Trans-Pacific Partnership (TPP) was being secretly negotiated, WikiLeaks obtained and published the proposed chapter on intellectual property. This was embarrassing to some of the governments, because they were seen as acting against the national interest, instead serving the interests of pharmaceutical and other companies. (This is not to mention that strengthening intellectual property provisions basically means restraining rather than freeing enterprise.) This is another example of how governments need to finesse the question of cover-up and exposure: this involves hiding the provisions and negotiations from citizens while telling everyone—especially politicians who have to approve the agreement—how wonderful life will be following approval. The idea is to obtain political backing without being influenced by popular resistance. (It should be noted that most politicians undoubtedly believe in the value of the trade deals they support.)

Next consider the value attached to trade agreements, positive or negative. As already mentioned, governments tout the advantages of the deals, appealing to the positive connotations of "free" in "free trade." Critics, on the other hand, have sought to discredit many trade deals, especially the ones mainly benefiting the rich and powerful.

Closely associated with values associated with trade deals are the many explanations of how they work or don't work. Proponents point to many advantages, usually ignoring harmful effects, and critics do the reverse. Critics often try to frame the deals as serving the interests of large corporations at the expense of national sovereignty, which nominally is under citizen control or at least influenced by citizens via elections and public debate. However, the responsiveness of elected representatives to the popular will is elusive when it comes to trade agreements, as indicated by the secrecy involved in the negotiations and the reluctance of governments to sponsor a wide-ranging public discussion.

The legitimacy of trade deals derives from their official status. They are inter-government agreements, and to the extent that governments have credibility, so then should the agreements. It would hardly seem fair if corporations simply stitched together a set of rules for trade and imposed it on the world's population. Govern-ments, especially those with fair elections, have much greater credibility for this purpose. Many members of the public trust what political leaders say, at least until blatant hypocrisies emerge: new leaders often have a honeymoon period, short or long, and may be able to push through the deals, especially when critics do not have details in advance to muster contrary arguments: many deals are *faits accomplis.* For corporations, governments are an essential part of the process to make the deal and to provide protection of private property and regulations to enable large corporations to thrive.

However, the fact that trade agreements are negotiated by governments provides remarkably little leverage for critics. This is the appearance of justice without the substance. After all, trade agreements are seldom a major election issue and politicians in any case do not have to follow through on election promises.

Finally, there are the methods of intimidation and reward. Intimidation of trade-deal critics through funding cuts or discrediting individuals is probably not as important as the great awards for those who support the deals. Some corporations and industry sectors receive direct financial benefits. Some individuals receive jobs and promotions. Journalists can gain better access when they are sympathetic.

In summary, governments play a double game in praising trade deals while restricting what members of the public know about the process and outcomes. Their basic strategy has to be to please two audiences: the corporations that benefit from the deals and the public that elects the politicians and which can agitate in opposition. The main ways that unpleasant truths about the deals—especially that their primary benefits are to large corporations—are dealt with is by hiding them from the public as long as possible. Meanwhile, the deals are touted under the assumption that benefits to the economy automatically lead to benefits for everyone.

Tax havens

Another source of tension for governments is the existence of tax havens, which are locations enabling individuals and businesses to avoid or minimise the tax they pay.[5] For example, a multinational corporation can establish its central office in a jurisdiction with low taxes and high financial secrecy, such as Switzerland, Hong Kong or the Cayman Islands, and then use transfer pricing to reduce its apparent income in higher-tax places like France and Sweden.

Tax havens are just one aspect of a wider process of tax avoidance and corrupt money transfers. Taxation is one of the powers of governments, and indeed one that enables the state to exist. Taxation can be thought of as an imposition on free exchange between individuals and groups; it is intended to be compulsory, and perhaps is the only state compulsion that remains widespread.[6]

In this context, it is not surprising that many people do what they can to reduce their tax, and many otherwise law-abiding citizens think nothing of cheating when it

---

5 Nicholas Shaxson, *Treasure Islands: Tax Havens and the Men Who Stole the World* (London: Bodley Head, 2011); Gabriel Zucman, *The Hidden Wealth of Nations: the Scourge of Tax Havens* (Chicago: University of Chicago Press, 2015).

6 Slavery and serfdom have been legally abolished, though forms still continue in parts of the world. Military conscription has been abolished in most countries, and jury duty and voting, though compulsory in some countries, are neither onerous nor difficult to avoid if really desired. Taxation, though, is standard everywhere. Only the means of imposing tax vary.

comes time to file their income tax forms. When trades-people ask to be paid in cash, it can be a sign that they do not intend to report the money as taxable income.

Although tax avoidance is widespread in many countries, the focus here is on the richest individuals and companies, the ones with annual incomes in the millions or billions of dollars. They have a capacity to pay, but commonly do what they can to reduce their tax bills. No surprise here. What might be surprising is that governments often seem quite happy to allow this to occur. They sometimes produce fiery rhetoric about tax avoidance but at the same time serve the rich at the expense of the poor, and this is something to be hidden when possible.

First, to take an extreme example of corrupt behaviour, consider loans to dictatorial regimes. In quite a few cases, the dictator and his family (very rarely her family) skim vast quantities of money from the loans into private bank accounts, held for example in Switzerland through a shell company in the Virgin Islands. Vast means billions of dollars. This is out-and-out theft. So what do Western governments do about it? They demand that the country honour the debt, namely that the corrupt government (or a successor government) cut government expenditure and raise taxes in order to pay interest and capital on the loans. Another approach would be to say to their own banks, "You made a bad loan. Too bad. You just lost the capital. Don't be foolish and do it again. If you want your money back, you'd better do something about Swiss banks that hide the proceeds of crime." In practice, Western governments usually allow these sorts of crimes to continue.

Then there is tax avoidance that is nominally legal. Large multinational corporations use transfer pricing to minimise their tax. This involves pricing internal transfers of goods and money within the company's operations in different countries in a way that ensures that tax is as low as possible. Usually this means that most of the profits appear to come from parts of the company based in low-tax places such as Ireland. In countries with higher taxes, it is seemingly miraculous that revenues of billions of dollars result in little or no profit.

If governments wanted to stamp out this sort of practice, it wouldn't be hard—at least in principle. After all, the rules for international finance are collectively made by governments and international bodies dominated by governments. In practice, corrupt practices and legal-but-unfair practices have continued for decades. The obvious explanation is that the most powerful governments operate to serve the wealthy and powerful at the expense of their own populations. This creates a challenge for governments: how to justify their policies to their own populations.

Consider possibilities for cover-up and exposure. I can speak of my impression of how this is dealt with in Australia: the role of tax havens and transfer pricing is seldom front-page news. It is more likely to be relegated to the business pages of some newspapers. Instead, governments encourage the media to report on cheating by those lower down, for example welfare fraud, when an unemployed person obtains more benefits than officially allowed. Low-level cheaters may be given stiff penalties, perhaps even going to prison, whereas executives of companies benefiting from massive rip-offs, legal or illegal, are seldom brought before a court.

There is a lot of reporting on taxation, with most of the attention on how taxes are too high, especially for high-income earners, with the explanation being that lower taxes are needed to offer incentives. However, tax evasion by rich individuals and companies only occasionally receives attention. There have been some scandals, for example the "bottom-of-the-harbour" schemes used to evade tax,[7] but these have not led to major reform. Official inquiries usually lead nowhere.

The following news report indicates the problem (the Coalition refers to the ruling Liberal-National Party government):

> Tax paid by companies controlled by Australia's richest business people, including Gina Rinehart, James Packer and Lindsay Fox, will remain secret after the Coalition succeeded in exempting private companies from new tax disclosure requirements.[8]

Australian billionaires found it embarrassing for information to be made public about how little tax they paid—sometimes almost none at all—so they quietly lobbied

---

7 Companies were stripped of their assets and profits and then, before taxes were due, transferred to new, poor owners. The stripped companies were metaphorically sunk to the bottom of the harbour.

8 Heath Aston, "Law change shields tax of wealthiest companies," *Sydney Morning Herald,* 16 October 2015, p. 4.

against the required disclosures. This illustrates how exposure can be a potent way of challenging injustice, and how governments can serve the interests of a wealthy minority at the expense of the Australian public.

It would be possible to examine additional methods to reduce outrage over tax havens and other forms of large-corporation tax evasion, under the categories devaluation, reinterpretation, official channels and intimidation. Only sometimes are these methods needed, because cover-up is usually adequate. Without going through a full gamut of methods, suffice it to say that governments play a dual game of stigmatising low-level tax evaders while avoiding giving attention to tax havens and high-level evaders.

Final comment

Economic inequality can be a source of public outrage, so government and corporate elites unite in dampening concern.[9] In relation to nationalism, there is a special challenge for state elites. By dint of their role in serving powerful groups, including those in other countries, they have a challenging task in maintaining the population's commitment to the country and to the state while reducing concern about inequality and actions that benefit the rich at the expense of others.

This is why corporate globalisation induces such a curious mixture of responses by governments, many of

---

9 Susan Engel and Brian Martin, "Challenging economic inequality: tactics and strategies," *Economic and Political Weekly,* Vol. 50, No. 49, 5 December 2015, pp. 42–48.

which promote or tolerate trade deals and tax havens that serve the global and mobile rich of the world at the expense of their own citizens who have less money and fewer options. Opposition to corporate globalisation can come from both ends of the political spectrum, from workers who feel threatened by cheap foreign labour, which can feed into racist feelings, and campaigners such as in the Occupy movement who challenge inequality. Examining the tactics used by governments provides a useful way of mapping the difficulties they face in reconciling nationalism and economic inequality.

12
The psychology of rule

"Who's the leader of your country? What do you think of him (or her)?" A few people will answer, "I don't know and I don't care." More commonly, though, people have strong emotional connections with rulers. These can be positive or negative. Quite a few liberal-minded US citizens had a visceral hatred of George W. Bush, while quite a few US conservatives detested Barack Obama.

Systems of rule are invariably accompanied by emotions and, more generally, psychological processes. Usually these facilitate the operation of the system.

Think of dictatorships in which the ruler is glorified. In China under the rule of Mao Tse-Tung, classrooms had several large photographs: Marx, Engels, Lenin, Stalin— and Mao himself. Think of the German Nazi regime with mass rallies, Hitler being the commanding figure.

Systems of representative government are not exempt from exalting the country's leader. In the United States, there is excessive attention to the president. Media speculation about the next president starts more than a year prior to an election: there seems to be more attention to the question of who is or will be the president than to policies. In other countries, a visit by the US president is a very big deal.

In countries with a monarch, even one without power, this provides a convenient figurehead that provides the basis for endless discussion. A royal wedding or the birth of a child in line for the throne receives great media attention, as if it makes any practical difference. But it does make a difference: it is part of the psychology of rule.

In parliamentary systems, citizens do not vote directly for the prime minister, who is chosen by elected party members. Gradually, though, prime ministers have taken on presidential attributes, so much so that opinion polls ask people their views about the prime minister and possible alternatives. The point here is that attention is constantly directed upwards, to the person at the top. In any moderately large country, few individuals ever have an extended interaction with the ruler. A photo opportunity perhaps, or a handshake, but in most cases the ruler is an icon, a figurehead, known through media coverage rather than personal contact.

A clue about the psychology of rule is the often-stated preference for a "strong leader," one who is decisive, commanding and leading the way, as the term "leader" might suggest. Strangely, though, this is in contrast with a leader who is cautious and consultative, which might seem to be more in tune with the ethos of democracy. Admiration for strong leaders may reflect a common pattern of treating leaders as rulers, admiring them for being dominant.

There is a body of research showing that people have a psychological predisposition to support the status quo or "the system," in other words the way the world is currently organised. John Jost and colleagues argue that, "there is a general (but not insurmountable) system justification motive to defend and justify the status quo and to bolster the legitimacy of the existing social order."[1] There is evidence that subordinate and oppressed groups may support the existing system as much as those in privileged and dominant positions.[2] It is possible that, after creating an egalitarian social order, this psychological motive might help to maintain support for it. However, in the present world order, system justification serves to encourage acceptance of the existence of governments, the state system and social inequality.

Insight into the psychological dynamics of rule is offered by gestalt therapist Philip Lichtenberg in his book *Community and Confluence*.[3] He draws on a standard idea

---

1 John T. Jost, Mahzarin R. Banaji and Brian A. Nosek, "A decade of system justification theory: accumulated evidence of conscious and unconscious bolstering of the status quo," *Political Psychology*, Vol. 25, No. 6, 2004, pp. 881–919.

2 This research has affinities with the moral foundation of authority, discussed in chapter 2.

3 Philip Lichtenberg, *Community and Confluence: Undoing the Clinch of Oppression* (Cleveland, OH: Gestalt Institute of Cleveland, 1994, 2nd edition). Other useful sources for understanding the psychology of rule include Arthur J. Deikman, *The Wrong Way Home: Uncovering the Patterns of Cult Behavior in American Society* (Boston: Beacon Press, 1990); Jeff Schmidt, *Disciplined Minds: A Critical Look at Salaried Professionals and the Soul-Battering System that Shapes their Lives* (Lanham, MD: Rowman & Littlefield, 2000); Judith Wyatt and Chauncey Hare, *Work Abuse: How to Recognize and Survive It* (Rochester, VT: Schenkman Books, 1997). There is a vast body of research rele-

in psychology: projection. In this process, a person disowns part of their own personality and attributes it to others, namely projects it onto them, rather like a movie projector puts an image on the screen. For example, a person who is often angry may complain about others being angry; a person who is forgetful may accuse others of forgetting things. A standard example is a man who is uncomfortable with the feminine side of his psyche, rejects it and sees it in homosexual men, who he detests or even attacks.

Lichtenberg says that projection dynamics are at play in attitudes towards leaders. Ordinary citizens forget or disown their own capacity to take initiative and instead attribute it to leaders. When citizens admire strong leaders, they disempower themselves (forget or reject their own capacities), project their own power onto the leader, and admire it.

For disliked leaders, the process is similar, just with a different emotional content: the key is not admiration or hatred for the leader, but the feeling that the leader has power and that the follower or subject does not.

Look to governments for action
The most obvious manifestation of this sort of projection is the expectation that for something to happen, governments need to take action, or perhaps stop taking action. The result is an incredible fixation on appealing to governments, through letters to politicians, petitions,

---

vant to the psychology of rule. The sources listed here are ones I have found useful from an activist and social change perspective.

meetings, and so forth. It's as if no one can act autonomously or independently: someone in power has to do the acting, and so if you want action, then get politicians or other government officials to do it.

I regularly see this with whistleblowers.[4] After they speak out in the public interest about corruption or hazards to the public, they are often subject to reprisals from bosses, senior management and, sometimes, co-workers. So what do they do next? They try to find some official body to take action to rectify the situation: the board of management, the ombudsman, auditor-general, a government inquiry, court or politician. At one level this makes sense: often the problems are far greater than what any one person can address. Power needs to be exerted. The question is, where does the power come from? Most whistleblowers instinctively look "upwards," towards those with more formal power, in government or government agencies.

An alternative source of power is found by looking sidewards, towards co-workers, ordinary citizens and action groups. To do this requires taking initiative, for example going to the media, going to meetings of campaigning groups, or helping organise a campaign. But many whistleblowers, and others subject to abuse and exploitation, feel they are so powerless that their only salvation is to find a saviour somewhere up within the system, a white knight who will come to the rescue.

---

4 Brian Martin, "Illusions of whistleblower protection," *UTS Law Review,* No. 5, 2003, pp. 119–130.

The process of projecting one's power onto leaders doesn't happen automatically. It is helped along in various ways, via education, media, elections and a psychological process called introjection.

Encouragement for projection onto leaders starts with what is taught in schools, including instruction (explicit or implicit) about the way the system is supposed to work: society, and especially government, is presented as a hierarchy, with some people in higher positions than others, and with those at the top making the crucial decisions. Relatively little attention is given to social movements and how ordinary people can organise and take action. Most schools are themselves organised hierarchically, with students being subordinate to teachers, teachers to principals, and perhaps principals to school boards or education departments. Students are taught to seek solutions to their own problems by going to teachers or the principal (or perhaps their parents), not to organise student protests.

The media are a major influence in encouraging people to project their power onto leaders. Media stories prioritise what governments do, both nationally and internationally. Politicians are regularly shown giving their views, in part because staffers seek favourable media coverage. Even without this, though, journalists and editors will run a story about the president or prime minister over one about grassroots action.

Media stories, as well as giving precedence to politicians and others with formal power in the system, also encourage projection by seldom providing any sense of how citizens can act on their own, without relying on leaders. There are some stories about trade unions, but usually about their actions, not about the daily slog of organising. There are some stories about environmental groups, usually with attention to spokespeople, not about what they spend most of their time doing.

The threat of global warming has triggered one of the world's greatest grassroots movements, with groups of all sorts taking action, talking to neighbours, cutting back on consumption, installing energy-efficient technologies and contributing to community initiatives. Yet to look at media treatments, nearly everything seems to depend on governments taking action. Governments do make a difference, to be sure. The point here is that media coverage encourages people to look to governments for solutions or to condemn governments for doing the wrong thing rather than suggesting how people can take action directly.

Then there are elections, in which candidates compete for people's votes in order to occupy leadership positions. The process of participating in an election can serve, in a psychological sense, as one of giving consent to the system of rule.[5] An unelected national leader can be seen as a dictator, as illegitimate; an elected national leader is legitimate and is a person to whom the population has willingly granted power. Of course not everyone votes and not everyone votes for the successful candidate, but still elections as formal processes of selecting leaders offer legitimacy and facilitate projection of power onto the

---

5 Benjamin Ginsberg, *The Consequences of Consent: Elections, Citizen Control and Popular Acquiescence* (Reading, MA: Addison-Wesley, 1982).

leader. After all, if voters have voluntarily chosen a leader, then deferring to that leader makes sense psychologically. Elections are a method of encouraging acquiescence.

This is one of the reasons that many dictators run sham elections. Even though nearly everyone recognises that the election has been rigged in one way or another, the process is a ritual that encourages acceptance of the outcome. In a way, it is analogous to singing the national anthem.

Education, media coverage and elections serve to encourage projection of power onto leaders, and leaders contribute to this through a psychological process called introjection. It involves, in this case, psychologically taking on the power of others. Leaders assume they have power, power that has been granted to them by their followers, subordinates or subjects. Now someone might say, "Well, actually, leaders *do* have power, so this thing called introjection isn't needed." This assumes the common model of power as something that powerful people possess and others have less of. However, a ruler does not exert power simply through what is in their own hands: their power depends on acquiescence or cooperation or eager support.

A military commander can do little if the troops refuse to obey. Arrest them and put them in prison! But this requires someone to do the arresting. Thinking about power this way leads to the perspective that it depends on quite a lot of people proceeding as if the ruler does indeed hold power as a possession: subordinates do as they are told, whether with enthusiasm or reluctance, knowing that if they don't, they may suffer penalties implemented by other subordinates who do what they are told. If all the subordinates got together and made their own decisions, the power of the ruler would evaporate.[6]

Introjection enables leaders to command more effectively. They believe, deep down, that a mandate has been granted to them, or that they are powerful, and the resulting feeling of authority helps them maintain the loyalty or acquiescence of others. In short, belief helps to maintain the reality.

When leaders deeply believe they are powerful, the corollary is that followers are relatively powerless. In practice, leaders can do little unless their followers support them, by doing their biding. Leaders, somewhere in their minds, may appreciate their own limited power, but to be effective commanders they have to get rid of this insight, so they project it onto their followers. The complementary process is that followers introject the belief of their own powerlessness projected by their leaders.

The concepts of projection and introjection are ways of understanding mental dynamics. If these concepts are not appealing, it may be more useful to talk about belief systems. Leaders adopt belief systems in which they are powerful and their followers are not, and many followers

---

6 The idea that people consent to being ruled was first articulated by Étienne de La Boétie, *Anti-dictator* (New York: Columbia University Press, [1548] 1942), with the title sometimes translated as *Discourse on Voluntary Servitude*. The trajectory of La Boétie's ideas has been examined by Roland Bleiker, *Popular Dissent, Human Agency and Global Politics* (Cambridge: Cambridge University Press, 2000).

adopt belief systems in which they are powerless compared to their leaders.

The processes of projection and introjection are most
obvious in the case of national leaders and power, but can
be observed elsewhere. Take for example the Nobel
prizes, bestowed annually on the person or group considered to have made superlative contributions to physics,
chemistry, physiology/medicine, economics, literature and
peace. When you stop to think about it, the committee
does not change the reality of a person's achievement. A
high-performing scientist does not suddenly have greater
achievement as a result of receiving a Nobel prize: their
achievement is the same; only the recognition has
changed. Yet many observers treat the awarding of a
Nobel prize as a type of anointment to greatness. Suddenly
the winner is highly sought after for interviews, talks, and
articles, and their opinions on all sorts of issues—in many
cases quite separate from their prize-winning research—
are treated with reverence. In psychological terms, greatness, in terms of brilliance and wisdom, is projected on
prize-winners, some of whom introject—psychologically
accept—this projection and start believing they are more
exceptional than before. (Of course many might already
have believed they are qualitatively different from others.)

Projection and introjection can be traced back to
other authority relationships, most obviously between
children and parents. It is apparent in the Stockholm
syndrome, in which captives, for example people who
have been kidnapped, start identifying with their captors
and lose the capacity to resist or escape even when the
opportunity arises. It relates to the idea of learned

helplessness: experiments show how mice, as a result of particular experiments, lose the capacity to try to escape electric shocks, even when the opportunity is at hand. Projection of power is also apparent in studies of obedience to authority, in which experimental subjects take actions, such as hurting another person, when instructed to by authority figures or simply encouraged to by the way the experimental situation is set up.[7]

Projection is easier when it is collective. If everyone else is applauding a political leader, it is easy to go along with the crowd. On the other hand, all it takes is a bit of dissent and it becomes easier to dissent.

Tactics of projection
Projection is a psychological state, orientation or process, and the focus here is on projection of people's power onto leaders, especially national leaders. To talk of the tactics of projection is to refer to methods that encourage this type of projection. These tactics follow directly from the previous discussion of the role of education, the media, elections and introjection in encouraging projection of power onto leaders.

First is *exposure* of the power of leaders, which is routinely highlighted in the media, especially during elections. Leaders themselves contribute through their interactions with others, often touting what they have accomplished, while seldom mentioning that they could do nothing without the governmental apparatus at their

---

7 Stanley Milgram, *Obedience to Authority* (New York: Harper & Row, 1974).

disposal. National leaders have media teams to promote their visibility, in a selective way, highlighting positives.

Second is *valuing* the power of leaders. Again, this is routinely promoted in schools, the media and elections. Of course, leadership is contested, so leaders are treated as good or bad depending on whether a voter supports them and/or their party. Still, the principle of leadership is seldom questioned. In schools or the media, there are few voices saying, "Maybe our national leaders should have less power."

Third is *explaining* that having powerful leaders is a good thing, or is just the way things are. The necessity of hierarchies is not often the subject of a careful analysis; it is more commonly assumed than argued. Arguments may be brought out in the face of criticisms. Otherwise they are usually relegated to academic journals. Least of all is the process of projection ever discussed.

Fourth is *endorsement* of leaders having power, and of citizens projecting their own power onto leaders. This occurs most obviously during elections, which can be understood as rituals in which voters endorse candidates, obviously enough, and more generally by participating endorse the system of electoral representation in which elected officials are granted power to make decisions on behalf of the rest of the population. Without the ritual, governmental power would not have the same legitimacy: elections serve a psychological purpose of encouraging projection of power onto leaders.

Fifth is *rewards* for projecting power onto leaders, and here it is possible to think of psychological rewards. Being part of a community with like-minded others is one reward: if everyone else is treating leaders as holders of power, then there is a satisfaction in conforming to this way of thinking. More deeply, projection of power allows relinquishing one's own agency and putting trust in a higher power. This can evoke the experience of childhood and trust in one's parents, something that for many can provide a feeling of security and safety. If the parent (national leader) is always there, is a source of good, and has been endorsed by the population, there is no need to assert oneself, namely to take the initiative to promote a different sort of society, one without powerful leaders at the top.

Tactics of counter-projection

One alternative to projecting power onto leaders is simply not to project it—not to put so much attention and expectations on leaders—but rather acknowledge one's own power to act, and assume the responsibility for doing what is possible in the circumstances. Another alternative is to project power to a collective, such as a trade union or activist group or social movement, while participating in it. These sorts of psychological alternatives, namely different ways of emotionally engaging with the world and the exercise of power in it, are systematically suppressed.

Cover-up is the first technique. Schools teach little about the agency of ordinary citizens compared to that of rulers; mass media give little attention to grassroots empowerment compared to the power of leaders; elections signal that the role of citizens is voting for rulers; and leaders, through their projection of their own dependency onto followers, discourage recognition of the capacity for autonomous action.

Devaluation is a second technique. In as much as grassroots, independent action is acknowledged as existing, it is typically painted as a threat or as ineffectual. Mass protests are portrayed as dangerous threats to the social order. For workers to demand decision-making roles in the production process is treated as subversion. And so on. The implication is that identifying with these manifestations of collective action is misguided, indeed almost a sign of mental disorder.

Reinterpretation is a third technique: it involves explanations of why psychological alternatives are wrong. Reinterpretation in other contexts, for example to justify shooting of peaceful protesters, can involve lying about what happened, minimising the consequences, blaming others, and framing the actions as legitimate. For psychological processes, these techniques are internalised within a person's thoughts and emotions. They can involve moral disengagement through processes such as displacement of responsibility, ignoring consequences, and dehumanisation.[8]

Official channels constitute a fourth technique for suppressing alternatives to projection of power onto

---

8 Samantha Reis and Brian Martin, "Psychological dynamics of outrage against injustice," *Peace Research: The Canadian Journal of Peace and Conflict Studies*, Vol. 40, No. 1, 2008, pp. 5–23. See especially the work of Albert Bandura, *Social Foundations of Thought and Action: A Social Cognitive Theory* (Englewood Cliffs, NJ: Prentice Hall, 1986), pp. 375–389.

leaders. Official channels include expert panels, ombuds-men, regulatory agencies and any other formal process that promises to provide justice. Elections are one important official channel. In the case of projection of power, official channels are the recipients of expectations for obtaining justice, and top-level leaders are the ultimate official channel. In psychological terms, the very existence of official channels creates the expectation that someone out there will be the savour who slashes through evil doings and provides salvation. By the same token, the existence of official channels discourages recognition that action can be taken directly, without relying on people in formal positions of authority.

Intimidation is a fifth technique for suppressing alter-natives. In the material world, this can involve threats, dismissal and physical attacks. In the psychological world, intimidation can occur by the threat of a different idea to a person's way of understanding the world and their place in it. One such threat is posed by cognitive dissonance, when ideas about the world clash with actual occurrences. Many people believe the world is just.[9] Poverty and exploitation pose a threat to this belief, and the solution can be the idea that people are to blame for their own misfortune, even when the evidence suggests otherwise. This is known as blaming the victim, and is a common phenomenon.[10] The idea that people have significant agency separately from

9 Melvin J. Lerner, *The Belief in a Just World: A Fundamental Delusion* (New York: Plenum, 1980).

10 William Ryan, *Blaming the Victim* (New York: Vintage, 1972).

leaders can be quite threatening, and promptly dismissed from consciousness. This is a sort of internal, psychological intimidation. It can be thought of as the process of introjecting powerlessness, which in practice means being fearful of one's own capacity to act.

Challenging the psychology of rule
The psychology of rule, including projection of power onto leaders and the introjection of powerlessness, can be deeply entrenched, sometimes deeper than actual rule. It might be said that, "You can take the ruler away from the people, but not the ruler out of their minds." After the execution of the king during the French revolution, it was not long before there was a new ruler, Napoleon; it might be that his rise was easier because of the population's long experience of being ruled. A similar dynamic occurred in Russia: after the overthrow of the oppressive rule by the Czar, the workers' and soldiers' soviets promised an egalitarian future but before long Stalin became dictator.

Many people assume that a person's personality is fixed, but actually personality is adaptable. Many people suffer from anxiety or depression or sometimes both. These are aspects of personality, and psychologists have spent enormous efforts in finding ways to change them. One of the most used methods is cognitive-behavioural therapy, in which a person learns to counter unwelcome thoughts by thinking about reasons why they are irrational. By doing this on a regular basis, it gradually becomes habitual, and levels of anxiety and depression can be reduced.

Some years ago I was a subject in a study of "personality coaching." Like other subjects, I first took the standard NEO Personality Inventory questionnaire, obtaining scores on the five main traits of personality, called neuroticism, extraversion, openness, agreeableness and conscientiousness. Each of these five areas has six sub-traits. For example, under neuroticism—more politely called emotionality—there are anxiety, anger-hostility, depression, self-consciousness, impulsiveness and vulnerability. After receiving our personality profiles, we received weekly coaching for a couple of months, with exercises to change any aspect of personality we chose. Many subjects decided to try to reduce their scores on a sub-trait of neuroticism, which makes sense: who wants to be anxious or depressed? I chose a different area: a sub-trait of openness called feelings, and over the period of the study my scores changed to reflect a greater receptivity to my own and others' feelings.

The point here is that personality traits, as normally measured, may be fairly stable, but they are not fixed. They are, in part, a response to environmental influences. If the traits of individuals can be shifted through coaching, it makes sense to think that traits of many individuals can be shifted by changes in culture and the economy. Quite a few observers of US culture have noted that narcissism—characterised by self-centredness, grandiosity, lack of empathy, and rage when prerogatives are threatened—has become far more common.[11] For example, surveys of

11 Jean M. Twenge and W. Keith Campbell, *The Narcissism Epidemic: Living in the Age of Entitlement* (New York: Free

college students show that in the matter of a few decades, far more see their goal in life as personal advancement, especially in making money, than serving worthy causes. Indeed, personal advancement is seen as a worthy cause! This increase in narcissism can be linked to the rise of neoliberalism and the associated promotion of materialism and individualism.

It also makes sense to think of personality as potentially malleable because of the many efforts to get people of think and behave in different ways. Some advertising is about encouraging people to buy particular products, but much advertising is about getting people to think of themselves in different ways, and in particular to be dissatisfied with themselves, as being incomplete and needing a product or service to fix the deficiency.

The psychology of rule is no different. There may be some basic tendencies in the human psyche, but the processes of projection and introjection can be changed, in two ways. One way is for people to project power to a different recipient; the other is to reduce the tendency to project power at all.

With this context, it is worth going through different types of tactics both to challenge the psychology of rule and to promote a different sort of thinking that might be called "empowered thinking." First is the tactic of exposure. To counter the constant attention to leaders in

Press, 2009). See also Sandy Hotchkiss, *Why Is It Always about You? The Seven Deadly Sins of Narcissism* (New York: Free Press, 2003); Anne Manne, *The Life of I: The New Culture of Narcissism* (Melbourne: Melbourne University Press, 2014).

education, the media and elections, it is not enough to highlight the bad aspects of individual leaders, because the deeper problem is the emphasis on leadership, at least with the assumption of hierarchy, with its formal differences in power. Hating leaders is not so very different from adoring them, because each involves projection of power. Perhaps being indifferent is a more suitable attitude to cultivate. To do this, avoiding attention to political leaders can be helpful, instead focusing attention on the power of so-called ordinary people.

The difficulty of doing this can be seen by trying to find textbooks that present history and politics from the point of view of the people rather than rulers. There are a few choices, such as E. P. Thompson's *The Making of the English Working Class* and Howard Zinn's *A People's History of the United States.* Even after cultivating a people's-history mentality, there is the challenge of every-day conversations. Within organisations, much gossip is about bosses, not about the capacities of co-workers, and then there is commentary on the latest news about local and national politics, nearly always driven by discussion about leaders. If you're regularly able to turn conversations away from politicians to how to work together independently of leaders, you have a rare skill indeed.

The next tactic is devaluing and valuing: devaluing the belief in the power of rulers and valuing the belief in the power of ordinary people. The devaluing of the power of rulers is a bit tricky. As noted earlier, it's not enough to be hostile to the current rulers, as that continues to assume that they are important, being worthy of investment of emotional energy. Turning love of a national leader into hatred of the national leader may make it easier to encourage challenges to this particular leader, but it is not clear whether this is a great improvement in challenging the emotional investment in *leadership*. Perhaps a more suitable goal is reducing or even removing the emotional energy invested in any leader, either positive or negative, and either current or future. The importance of this can be seen by noticing how many people who detest a current leader pin great hopes on some future one. If salvation is seen as coming from a change in leadership, the projection of power onto leaders has not been devalued.

Perhaps a better attitude is indifference, ignoring the constant media coverage and discussions about national politics (or paying little attention to speculations about what the boss will do, or who will be the next boss), or perhaps treating all this attention with an attitude of detached amusement, rather the way you might respond to attention to a celebrity about whom you have little knowledge and no interest. How to foster such an indifference or detachment is a big topic. At an individual level, it might mean reducing media consumption. At an interpersonal level, when talking with friends for example, it might involve switching the topic or developing some humorous gibes about the constant attention to leaders. With some friends, it might be possible to say, "It's fascinating how the prime minister has been able to entice you into paying attention to herself/himself." With others, "It's really boring to talk about the prime minister." Or, "Aren't there some other people we could talk about?"

Depending on your occupation and position, you might have a more direct way to influence the valuing of others. As a journalist or blogger, you can make choices about focusing on leaders and their agency, for example focusing on government policy, or on citizens and their agency, for example local initiatives for change. As a manager, you can make choices about how to interact with subordinates, either as a director or a facilitator; to foster agency by your subordinates, you can try to avoid introjecting power and deflect others' interest in your thinking and instead encourage independent thinking, for example by nominating a person to be a devil's advocate. In some techniques, there's a fair bit going on besides valuing. The point is that by changing one's behaviour and fostering behaviour change in others, it's possible to influence their ways of feeling about power and agency.

The next tactic is interpretation, which means explaining what's going on. In this case, interpretation is about the ways of explaining the distribution of power. Interpretation tactics that serve rulers involve explaining unequal power as natural, inevitable, functional, necessary or unquestionable. To challenge such interpretation tactics, alternative views can be presented that leaders are power-hungry, self-serving, corrupt and a danger to society and that it is much better to develop the capacity of ordinary people to cooperate and make decisions for themselves. In short, rulers are not needed.

There is plenty of writing and examples available that can be used to counter the standard interpretation techniques, and which can be introduced in conversations, meetings, blogs and campaigns. How much this can shape feelings about rule, in particular the projection of power onto leaders, is an intriguing question. If people were entirely rational, then arguments and evidence would be sufficient to change thinking and behaviour, but people are commonly driven by their intuitive minds.[12] Projection of power is hardly ever the result of a calm, careful analysis of desirable ways of emotionally relating to rulers and subjects. Likewise, overcoming projection of power is seldom going to be achieved by arguments alone. Nearly always, experience—for example, involvement in grassroots campaigns—is more likely to influence gut reactions. After gut reactions shift, then a person may seek out evidence and arguments to support their new intuitive feelings. So evidence and arguments are valuable, but more to support those who already have corresponding feelings than to create those feelings.

The fourth set of tactics is discrediting tactics used by rulers and endorsing alternatives. Translated into the psychology of rule, this means discrediting projection of power onto rulers and instead endorsing accepting one's own power and capacity to act.

It's worth reiterating that discrediting rulers' tactics does not necessarily mean discrediting particular rulers. After all, lots of people hate the president, or the boss for that matter. To hate a person is still to invest emotional energy in them, and usually to project some power onto them. Lichtenberg observes that *agents* of rulers, such as police, soldiers and informers, often are psychologically fused with rulers. When those who are weaker develop a passionate hatred of these agents, such as activists who detest the police, this can reflect a projection of their own

---

12 See chapter 2.

tendencies to identify with rulers. In other words, scorning, blaming or hating the agents is a means of warding off a desire to submit to power. Lichtenberg recommends that challengers learn about their own psychological tendencies by interacting with agents of power.[13]

Rather than condemning agents of power, what should be involved here is discrediting *rulership,* namely the structures and processes of domination, including the benign exercise of power and control. It might be easy to reject domination at an intellectual level. What's needed is changing one's intuitive response, to react at a gut level against rulership, and favourably towards non-hierarchical alternatives.

There is research showing that people's reactions to sexually or racially coded information—for example pictures of people—are deeply embedded in their minds. You might think you aren't prejudiced, but sophisticated experiments show that most people react differently in their brains to images of men and women, or black and white people.[14]

One way to change automatic responses is to practise by using conscious attention and behaviour to shape intuition. An example is for a shy person to pretend to be outgoing, for instance to approach strangers and start a conversation. At first it feels uncomfortable, because the intuitive mind yells out in pain. After a few months of

---

13 Lichtenberg, *Community and Confluence,* 91–95.

14 Mahzarin R. Banaji and Anthony G. Greenwald, *Blindspot: Hidden Biases of Good People* (New York: Delacorte Press, 2013).

practising being outgoing, the intuitive mind learns from the actual behaviour that it's okay, and stops rebelling.

How to apply the same approach to challenging the automatic projection of power remains to be systematically tested. It's plausible to think it can occur by regularly associating revolting things, like a detested food, with systems of domination. Likewise, a parallel process of valuing alternatives to rulership could be developed.

The fifth and final set of tactics involves rewards, either refusing the rewards provided by leaders and fostering and accepting the rewards of equal relationships. In the case of the psychology of rule, the rewards are psychological rather than being money, power or position, but psychological rewards can be just as potent as any others.

The reward from projecting power onto rulers is being freed of any expectation of agency or responsibility. It is like becoming a child who trusts parents to protect them. It is a feeling of security. Projecting power can provide a psychological reward even when the parent/leader is oppressive, because this still means acquiescing and not being burdened with the expectation of escaping or challenging the ruler and acting autonomously.

The tactic of rulers is to encourage projection of power, and to introject power, so the counter-tactic is to refuse to project power. This means accepting one's own power, not relying on rulers or leaders or bosses to be the solution to problems, but instead thinking, planning and acting in whatever way is possible. It means taking direct action rather than appealing to leaders to take action. It means planting a community garden rather than asking for official permission to set up a garden. It means using encryption and other techniques for secure communication rather than relying on government agencies to protect privacy. It means cutting your own greenhouse gas emissions or joining the "transition town" movement for energy security rather than appealing to national leaders to establish policies to deal with climate change. It means helping communities prepare to defend against aggression rather than relying on military defence.[15]

These examples also point to the parallel process of providing rewards for alternatives. The psychological rewards from direct action include the satisfaction of exerting one's own agency, of making practical steps towards alternatives, and of working with others in a common cause. Setting goals and working with others towards achieving them is known to improve wellbeing.[16] Psychologically, reducing projection of power and taking on more responsibility for one's future can be satisfying indeed. This satisfaction can be the basis for continued efforts to overcome projection of power and build a society without domination.

Conclusion

To challenge systems of domination, action is crucial, and there is plenty of effort put into methods such as protests, strikes, boycotts, setting up alternative systems of govern-

---

15 Brian Martin, *Social Defence, Social Change* (London: Freedom Press, 1993).

16 Sonja Lyubomirsky, *The How of Happiness* (New York: Penguin, 2007).

ance—and armed struggle, too. Taking action is essential, but it does not always lead to changes in the way people think and feel. If people feel more secure when projecting power onto leaders, then overthrowing a repressive government may simply be the prelude to another autocratic ruler.

One way to foster a psychology of autonomy, self-efficacy and cooperative endeavour is to begin behaving towards others in ways that reflect these ideals. This can be done in campaigning groups and in day-to-day interactions. By behaving in egalitarian ways, gradually the psychology of rule is transformed into a psychology of egalitarianism, along the lines of the sayings "Be the change you want to see" and "Live the revolution." These slogans contain important truths: change starts now rather than after the revolution, and personal change is part and parcel of social change. By following the sentiment in these slogans, there is another process, or rather set of processes: changes in behaviour lead to changes in thought and emotion, and vice versa.

While changing the psychology of rule via new modes of action is vital, there is also a place for a direct focus on psychology, in particular on the mutual processes of projection and introjection of power. In this chapter, the focus has been on tactics by which projection is fostered and challenged. Usually, when thinking about tactics, they are out in the world of action, in business, military or activist campaigns. But struggles over the way people think and feel can also be thought of in terms of tactics, and the same sorts of tactics are relevant as in other domains: exposure, valuing, interpreting, endorsing and rewarding, and their opposites.

One of the advantages of focusing on psychological tactics is that it is possible to begin immediately. There is no need to join an action group (though that might be helpful) and formulate a campaign strategy. Anyone can start observing their own environment—including media consumption, everyday conversations, topics that trigger emotions, and sensations of discomfort and relief—and experimenting with different ways of talking and thinking. It may not seem like doing a lot, but it can be part of a wider process. It is important, too. Otherwise, why would there be such incessant efforts to encourage people to project power onto leaders?

Finally, there is much to learn about the psychology of rule and of egalitarianism. These are not important research topics in psychology, nor do activist groups systematically develop ways of changing the ways people think. Indeed, many activists see salvation in different rulers, or in their own activist leaders, rather than in alternatives to rulership itself. Of course, there is plenty to debate in this area, and not everyone aspires to end expectations about dependence on leaders. What is important is to openly address the issues of leadership, rule, projection and introjection.

13
War

Just after World War I, US essayist Randolph Bourne wrote, "War is the health of the state." This statement captures key insights about patriotism: war is a means of both strengthening state power and stimulating loyalty to the state.

An ultimate test of loyalty is willingness to die for one's group. The key question is, "what group are you willing to die for?" Some parents are willing to die for their children. But why should young men be eager to risk their lives for an abstract entity called a country? That is a mystery. An even stronger test of loyalty is willingness to kill for one's group. Why should anyone offer to kill a stranger on behalf of an abstraction?

At a general level, war functions to accentuate group identification. There is a threat to the group, so members rally in defence. The threat is from the "enemy": to safeguard the group, the enemy must be defeated, even destroyed. This impulse is deeply rooted in human evolution. But this still doesn't explain why such strong loyalty can be attached to the country and government rather than to some other entity, such as the family. After all, in modern warfare, defeat does not necessarily mean destruction for families or individuals—just a new set of rulers, perhaps more benevolent ones. Why would a mother or father expect a son to risk his life for a country?

Part of the answer is that governments use a number of techniques to foster identification and loyalty.

In Europe in the late 1800s, the socialist movement gained great strength. It was epitomised by the slogan "Working people of the world, unite!"—though in practice the actual slogan referred to working men, with women left out of the picture. The idea was that the working class would stand together against the ruling class. As political crises hit Europe in the early 1900s, with the possibility of war, socialist leaders called on workers to refuse to fight each other. But then came the so-called Great War beginning in 1914—today called World War I but perhaps more accurately called a European war—and most workers rallied not against the ruling class but in support of their governments, to fight and kill each other, sacrificing their lives for their states. This was the context in which Randolph Bourne said that war is the health of the state. World War I stimulated patriotism, strengthened European states against their own populations, and undermined hopes of a peaceful transition to socialism.

In his famous novel *1984,* George Orwell envisaged a world divided into three competing superstates, Oceania, Eurasia and Eastasia, constantly at war with each other. War provided the pretext for dictatorship, including pervasive surveillance of citizens, including the novel's protagonist, Winston Smith. The novel was completed in 1948, and it can be argued that Orwell was portraying not a future dystopia but rather elements of contemporary reality, in the Soviet Union and other repressive communist states of the time as well as aspects of so-called western democracies, just emerging from years of total warfare in which citizens were subordinated to the common struggle against the enemy, and about to plunge into a struggle called the cold war in which there was the potential of destruction by nuclear weapons.

PROMOTING PATRIOTISM

Efforts to promote patriotism are especially prominent in relation to wars. To illustrate some of the methods used, I will use a range of examples, especially from World Wars I and II, which involved unprecedented mobilisation of societies for war.

Exposure

A crucial technique is *exposure*: war receives high visibility. Governments naturally want to highlight their efforts against the enemy. The mass media, with their preoccupation with conflict and emphasis on proximity and local relevance, give saturation coverage of war-related stories. During wartime, governments and mass media operate together to highlight relevant issues, for example that sacrifices are needed, that resources for war-fighting are top priority and that troops are putting their lives on the line.

Valuing

Exposure usually operates in conjunction with *valuing*: the war effort is seen as worthy. Supporting the government is patriotic. Troops are glorified. This can occur in the media, but is even more potent within families and local communities. In Australia during World War I, men who volunteered for the army were seen by many as brave, loyal and indeed everything a man should be. For many women, a man in a uniform was far more desirable than one not in the military. Supporting the troops became a test of loyalty.

The glorification of troops continues after wars are over. After World War I, monuments were constructed throughout Australia in memory of the soldiers who died in the war. In Canberra, the national capital, the War Memorial is an impressive building with the name of every Australian soldier who died in any war engraved on a wall. In small towns and local suburbs throughout the country, there are smaller memorials to soldiers.

This glorification of Australian soldiers occurred despite the fact that Australia was not even under attack in World War I: soldiers were sent to Europe to fight on behalf of Britain, the home country. Australia had been a British colony, only becoming an independent country in 1901. So Australian nationalism was subordinated to British agendas.

The glorification of Australian soldiers occurred despite World War I being a massive sacrifice of lives for little purpose. Anzac Day, 25 April, is an Australian public holiday in honour of military personnel who served in wars. Anzac stands for the Australian and New Zealand Army Corps. The year 2015 was the one hundredth anniversary of the landing of Australian and New Zealand soldiers at Gallipoli, in Turkey, where they futilely tried to advance against Turkish troops. A bloodbath resulted, with high casualties on both sides.

Even supporters of the war might say that this episode in Australian history was an absurd waste of lives and that British commanders were incompetent. Furthermore, some Australian soldiers at the time said they respected their Turkish counterparts. Yet the overwhelming sentiment remains that these Australian soldiers were brave, advancing in the face of almost certain death. Sacrificing their lives for their country was noble. All those who "served their country" in uniform are honoured today, but especially those who lost their lives in battle. Death is thought to have brought them a type of greatness.

Critics of war might harbour different thoughts, for example that these soldiers were naive and foolish pawns in an insane, purposeless conflict, that they would have been braver to have not joined the army, or that as members of the working class they should have been fighting against their upper-class commanders rather than other working men. But such thoughts usually remain private. Articulating them in public is to transgress against a ritual that retains the full endorsement of the political establishment.

Explanation
A third technique to promote patriotism in relation to war is *explanation,* namely providing plausible reasons why military defence is necessary. In many cases, formal explanations are not needed, because of underlying assumptions: there is an enemy, actually or potentially dangerous, and the threat must be countered by lethal force. Note that there are several assumptions involved in this seemingly simple proposition: (1) there is an oppo-

nent; (2) the opponent is dangerous: an enemy; (3) the way to counter this dangerous enemy is through military means.

The first assumption—there is an enemy—appeals to the idea that *we* are a group and *they* are not part of the group, and hence *they* are an enemy. The essence of fostering patriotism is the ensure that the in-group is thought of as the country or state or nation, and not some other grouping such as an extended family, business, sporting club, social class or network of like-minded individuals.

The second assumption, that the opponent is dangerous, grows out of a common expectation that out-groups are a threat to the in-group. An alternative is that the out-group is actually more desirable. Maybe the so-called enemy is actually a friend bringing salvation. This, to a patriot, is treason, discussed later. For the purposes here, the assumption of an enemy is part of the rationale for the military.

The third assumption—that military defence is necessary to counter the dangerous enemy—builds on the common belief that the only way to oppose violence is through superior violence. Defenders of military defence hardly need to argue that the only way to stop an invasion is through military means.

The rationale for military forces can sometimes require dubious logic. A classic example is the theory of nuclear deterrence touted during the cold war. From the side of the US government and its allies, the Soviet bloc was the enemy; it was dangerous because of its armed forces, especially its nuclear weapons; and the only way to counter this threat was through superior force, including a superior nuclear arsenal. The Soviet government was told that if they attacked, they would be met by an overwhelming counter-attack, destroying them. This threat was supposed to deter them from attacking. The Soviets were assumed to think in exactly the same way, so the result was deterrence via mutually assured destruction or MAD.

This rationale contained several flaws. Because of secrecy about the capability of nuclear arsenals, it was easy to exaggerate the threat. In the 1960 election campaign in the US, John Kennedy campaigned on a claim that there was a "missile gap," namely that the Soviet nuclear arsenal contained more missiles, even after being informed by military figures that no such gap existed.[1] In fact, the US nuclear arsenal was far superior, so it was the Soviet missile forces that suffered from inferiority. Threat exaggeration has been a recurrent feature of US strategic nuclear policy-making.

Another flaw in the doctrine of nuclear deterrence is its selective application, which operates with thinking like this: "It's good for us to be strong to deter the enemy, but some enemies are so dangerous they should not be allowed to deter us." In the 1970s, most of the world's governments signed the Nuclear Non-Proliferation Treaty. The governments of existing nuclear weapons states—US, Soviet Union, Britain, France, China—pledged to reduce their arsenals, while other governments pledged not to

---

1 Gary A. Donaldson, *The First Modern Campaign: Kennedy, Nixon, and the Election of 1960* (Lanham, MD: Rowman & Littlefield, 2007), p. 128.

acquire nuclear weapons. The idea of the treaty was to stop "proliferation" of nuclear weapons capabilities, namely to stop additional governments getting their own arsenals. But what does this say about the doctrine of deterrence? If governments are deterred from attacking by nuclear weapons in the hands of enemies, then surely more governments should have their own arsenals, and eventually military aggression, or at least nuclear aggression, would cease.

The double standard in reactions to nuclear weapons arsenals is sometimes acute. The US government has repeatedly raised the alarm about weapons programmes in other countries, notably North Korea, Iraq and Iran, all the time sitting on its own arsenal of thousands of nuclear weapons with sophisticated delivery mechanisms. The US government claims it needs the weapons to deter attackers, but desperately wants to stop other governments acquiring their own deterrents. The 2003 invasion of Iraq was launched on the pretext of stopping the threat of Iraqi nuclear weapons, a threat that turned out to be non-existent.

Then there is the case of Israeli nuclear weapons, an arsenal thought to number dozens or hundreds, about which US policy makers never raise any concern. The implication is that deterrence doctrine involves an implicit double standard: nuclear weapons are a deterrent, or just not even mentioned, when they are in the hands of the good guys, but are a grave threat to world peace when in the hands of bad guys.

The case of nuclear weapons and deterrence theory is just one example of the rationale behind military races.

The enemy's military threat is misperceived, almost always by being exaggerated, thereby justifying a military build-up that is seen as entirely defensive and used to maintain peace. In blunt terms, *our* military is for peace, *theirs* is for war. Deterrence theory and related logical-sounding rationalisations serve to hide or sugar-coat this basic assumption.

Another common explanation of the need for military force is to defend against attack. However, in many cases there is no credible threat, yet threats are still invoked. One of the arguments is that a threat may arise suddenly, so military preparedness is required just in case. Think of New Zealand, thousands of kilometres away from other major population centres and of no strategic significance. Yet the government of New Zealand maintains military forces, allied to the US government.[2] The argument about the need for defence is plausible when there actually is a threat, but when there is no threat but no major reduction in military preparedness, this exposes the argument as hollow.

Endorsements
Another key method of promoting group loyalty to the state and its military forces is *endorsement*. In most countries, nearly all prominent individuals—politicians,

---

2 The New Zealand government is not as tied to the US military as the governments of Australia, Britain or Canada. For example, in the 1980s the New Zealand government refused to allow visits of US nuclear ships, much to the annoyance of US political leaders.

religious leaders, business executives, heads of government departments, and others—endorse the troops. There may be disagreements about particular wars, weapons systems or levels of military expenditure, but very few people of significance question the basics about military forces. To the contrary, many of them state their commitment: supporting the military is a test of loyalty, to the extent that anyone who is seen as too weak in their enthusiasm may be accused of being unpatriotic.

Rewards

Rewards are another method of promoting patriotism in relation to war. In Palestine, Hamas provides financial support to families of suicide bombers. To some, this is outrageous, but most other governments give extra benefits to at least some of those involved in war-making. Veterans may have special hospitals and medical services, and may receive special pensions. In the US after World War II, the GI Bill gave veterans special access to higher education. Many veterans and their families say not enough is done for those who risk their lives on behalf of their countries. However, many others commit their lives to helping others—nurses, teachers and fire-fighters, for example—but do not receive special benefits.

Far more than material benefits are the psychological rewards, with soldiers being treated as heroes. Some who display special valour receive citations.

Then there are the rewards for those at the top of the hierarchy: commanders, generals and top politicians. Wartime leaders who perform well are commonly seen as exceptional individuals and greatly admired. A classic example is Winston Churchill, Britain's prime minister during World War II. Outside of this war, his record was far less noteworthy. The cult of the leader is found in many dictatorships; war, requiring mobilisation of a society to defend against the enemy, exalts leaders even in systems of representative government. This is because uniting in a cause encourages individuals to put their trust in the leader, and project their own sense of agency to the leader.[3]

National leaders thus have much to gain from fostering conflict. An enemy is, in a sense, a leader's ally in building support for the state.

In summary, there are five main ways to promote patriotism and state-centred thinking in relation to war: exposure, valuing, explanations, endorsements and rewards. When these work effectively, they become part of the culture, adopted by individuals as part of their thinking and overriding other loyalties. This is most dramatically demonstrated when individuals are willing both to kill and to sacrifice their lives for their country and when family members are proud they have done so.

CHALLENGES

Not everyone goes along with the glorification of war and the patriotic duty to support the state against its alleged enemies. Indeed, in many places opposition to war has been vociferous and sustained. There is nothing natural in war-related patriotism: support for the country, and for its

---

3 See chapter 12.

military forces, is only one way in which loyalty can be assigned. The existence of alternative loyalties is why continued efforts are exerted to promote patriotism and to hide or discredit alternatives.

The next step in analysing tactics of patriotism in relation to war is to examine direct challenges, taken separately from promoting alternatives to war, which I address later. Each of the five main methods of promoting patriotism can be countered. This is a huge topic. For example, peace movements have used a wide variety of methods, including advertisements, petitions, rallies, marches, refusal to join the military, and blockades. Many of these actions are in relation to particular wars or weapons systems, for example nuclear weapons.

Only some of these challenges to war present them-selves as direct challenges to patriotism. Indeed, some peace activists are careful to portray themselves as true patriots, serving their country's interests by opposing disastrous policies that lead to death, destruction and loss of civil liberties. Furthermore, peace activists are often quite respectful of the troops, emphasising that their opposition is to policies and practices, not individuals. In this section, I present a few examples of challenges that more directly target the promotion of patriotism in relation to war. Many of these confrontations involve presenting alternatives to war, for example diplomacy or nonviolent action; I will address these later.

Challenging pro-military messages
First consider the high visibility of war stories, war reporting and war memorials. Many challenges to the

exposure of war occur out of sight. For example, a local government might be planning to build a memorial to war dead, and some staff members argue that the funds could better be spent elsewhere, or that a memorial be built in honour of peace campaigners. Librarians might choose to order books on peace rather than war. Panels in charge of the syllabus for a school district might prefer a text that gives less prominence to war. There are many such quiet battles over the visibility of war.

Most reporting on conflicts gives a one-sided perspective, with emphasis on violent acts and on simplistic storylines involving good guys and bad guys. Watching the news, it is very hard for viewers to appreciate the sources of conflict, to understand the complexities involved, or realise that nonviolent methods are being used. For example, news about the Israel-Palestine conflict seldom gives any indication that nonviolent methods— such as protests, strikes, boycotts and occupations—are regularly used.

Critics of this usual approach to reporting conflicts have called it "war journalism" and have proposed an alternative, "peace journalism."[4] It involves offering a broader, more in-depth treatment of conflicts, including driving forces, historical context, different participants, options for resolution, long-term impacts and so forth. To the extent that journalists—both professionals and citizens—take up the principles of peace journalism, reporting of conflicts is transformed: a different sort of picture is

---

4 Jake Lynch and Annabel McGoldrick, *Peace Journalism* (Stroud, UK: Hawthorn Press, 2005).

presented, with less emphasis on the latest violent clash and more information about causes, motivations, multiple players, precedents, initiatives, options and solutions. Peace journalists, rather than racing to the scene of some new atrocity, will be investigating ongoing conflicts—often ones invisible in war journalism—probing the back stories and exposing dimensions normally ignored.[5]

Devaluing war and the state

Given that glorification of troops and their noble cause is standard in the usual war-linked patriotism, one option for challenging war and the state is to do the opposite: treat them as misguided, worthless, counterproductive, reprehensible or criminal. This is risky territory for opponents of war, because defenders of the faith are very sensitive to any criticism—especially criticism of soldiers.

On Anzac day, 25 April, in all parts of Australia there is a dawn service to remember soldiers who lost their lives in war, and a march in which veterans participate, some wearing their uniforms. The annual Anzac Day march is not a promising time to challenge any part of the Anzac legend. In 1980 in Canberra, the national capital of Australia, a group of women attempted to join the Anzac Day march in memory of women raped in war.[6] They carried placards including "Rape is war against women," "Soldiers are phallic murderers" and "Women are always

5 Virgil Hawkins, *Stealth Conflicts: How the World's Worst Violence Is Ignored* (Aldershot, UK: Ashgate, 2008).

6 This information is drawn from articles and letters in the *Canberra Times*. Copies available on request.

the victims." They planned to lay a wreath with a sign saying, "In memory of women raped in war." This protest action was a direct challenge to the mythology of the noble Anzacs: it suggested that some of them might have been rapists. It is well documented that rape by soldiers is a frequent occurrence: women in conquered territories are prime targets. Sometimes rape is a conscious tool for subjugating populations; more often it is an act in which men take advantage of their power and the absence of any policing of their crimes.

Police arrested 14 women, alleging there was an imminent breach of the peace. (It is ironic when protesters against war are charged with breaching the "peace.") In September, a special magistrate convicted the women. Most received fines; three were jailed for a month. According to a newspaper story, the magistrate said they were "social mutineers" who were involved in "wilful and collective defiance of authority, of a sort which in a military sense would be called mutiny." The three who were jailed were said to have a "tendency to become social anarchists."

The attitude of the police and the magistrate—shared by many of the veterans marching on Anzac Day—reflects an extreme antipathy towards any action that devalues soldiers, in this case by pointing to actions by soldiers that are usually ignored in remembrances of a glorious past. It is unthinkable that the troops were anything less than noble.[7]

---

7 The magistrate's comments stimulated a storm of protest. Dozens of women prepared for civil disobedience at the following

In many parts of the world, it remains risky to show disrespect towards veterans. Lindsay Stone discovered this the hard way. She liked to take photos of herself making provocative irreverent gestures, as a way of having fun. One photo she posted on social media was of herself making a rude gesture in front of a military cemetery. This was taken up by critics, and Stone was inundated with hundreds of thousands of abusive comments. As a result, she lost her job.[8] This illustrates that many people continue to be very upset by anyone showing disrespect for soldiers. It also suggests that challenging the glorification of troops is risky.

It is far safer to criticise political leaders who take countries to war. The troops, after all, are just doing their jobs.

With the abolition of conscription in many countries and the rise of professional armies that use economic incentives for recruitment, is it safer to challenge the reverence associated with being a soldier? Professionals are volunteers, to be sure, but no longer in a sacrificial mission as in World War I. There are many others who volunteer for dangerous occupations, such as fire fighting and coal mining. Furthermore, the risk to many members of military forces in western armies is minimal. Those who sit in bunkers in Nevada and pilot drones on the other

year's Anzac Day march. Meanwhile, the government passed a new law against such protests. In the end, hundreds of women were allowed to join the march.

8 Jon Ronson, *So You've Been Publicly Shamed* (London: Picador, 2015).

side of the world are not risking their lives, though their jobs require skill and dedication.

Then there are mercenaries, a category of soldier different from volunteers or conscripts: mercenaries are soldiers for hire. In the US, mercenaries are called contractors, a euphemism. Rather than being front-line soldiers, contractors more commonly fill support roles such as driving vehicles, and undertake unsavoury operations such as interrogations, renditions and assassinations. Few members of the public realise that in the Iraq war beginning in 2003, there eventually were more US contractors than US troops. Though most contractors are highly professional and motivated by wanting to help others, nevertheless to be seen as a "gun for hire" is not nearly as glorious as being a regular soldier. So it is not surprising that the US government plays down the role of contractors and emphasises the contribution of its regular armed forces.

In many wars, some politicians and soldiers are guilty of war crimes. This might be waging an unjust war, killing civilians, torturing enemy troops and committing or tolerating atrocities. Exposing these crimes is a powerful way to discredit those involved.

After World War II, leading Nazis were charged with war crimes and brought to trial in Nuremberg, Germany. This was a more civilised way of addressing war crimes than the more common approach of summary execution. Nevertheless, what is striking about responses to war crimes is that nearly always it is the enemy that is targeted. Making a case that the victor, or the more powerful side, was guilty of war crimes is a potent way to discredit war-makers, but it is difficult to get many people to pay attention. During World War II, the Allies carried out extensive bombing of civilian targets in Germany and Japan, yet few called this a war crime.[9]

Challenging justifications for war

Part of the connection between war and patriotism lies in the official justifications for going to war and continuing in war. Challenging the official rationales thus plays a role in challenging the patriotism-war link. Doing this is an important task, and one often done extremely well. There are numerous speeches, articles and books that question particular wars, or war in general, with careful arguments and ethical considerations.

Prior to the US-government-led invasion of Iraq in 2003, there was a massive protest movement. As part of this movement, various writers challenged the official rationales for the war. After the invasion, the intellectual questioning of the enterprise continued.[10] However, this level of questioning is unusual. US military involvement

---

9 Eric Markusen and David Kopf, *The Holocaust and Strategic Bombing: Genocide and Total War in the Twentieth Century* (Boulder, CO: Westview, 1995).

10 See for example Michael Isikoff and David Corn, *Hubris: The Inside Story of Spin, Scandal, and the Selling of the Iraq War* (New York: Broadway Books, 2007); Sheldon Rampton and John Stauber, *Weapons of Mass Deception: The Uses of Propaganda in Bush's War on Iraq* (New York: Tarcher/Penguin, 2003); Norman Solomon, *War Made Easy: How Presidents and Pundits Keep Spinning Us to Death* (New York: Wiley, 2005).

in Vietnam began in the 1940s with support for French colonialists, and continued through the 1950s and 1960s. The US movement against the war gradually developed in the 1960s, along with the escalation of the war itself. Noam Chomsky's trenchant criticisms of US policy, for example in *American Power and the New Mandarins,* played a significant role in stimulating opposition.

Going back to earlier wars, well-articulated opposition sometimes took quite some time to develop. More important, in many countries, was the fact that governments suppressed criticism. In Nazi Germany, there might have been critiques of Hitler's war plans, but they did not have a high public profile.

Challenging justifications for war can also be done retrospectively, in histories. Very few histories of the US offer comprehensive critiques of the war of 1812 or the Mexican war, for example.[11] Challenging pro-war and one-sided histories is important in countering the usual justifications for war.

Challenging endorsements
When national leaders and other high-profile figures say they support greater military expenditures and greater preparedness for war, this gives greater legitimacy to the military and the state. Many people do not examine the

---

11 The classic source is Howard Zinn, *A People's History of the United States* (New York: Harper & Row, 1980). See also Mark Cronlund Anderson, *Holy War: Cowboys, Indians, and 9/11s* (Regina, Saskatchewan: University of Regina Press, 2016).

arguments themselves, but rather base their views on those in authority or who they respect.

There are several ways to counter endorsements. One option is counter-endorsements: find some prominent individuals who will make statements challenging the military. If they are military figures, it's even more effective.[12] Just a few counter-endorsements can be effective, especially when they change a monopoly of elite opinion in a contested domain. This can make some people unsure of what they should think.

Another approach is to expose something wrong with those making the endorsements. Perhaps they have made rash or inaccurate claims in the past. Perhaps they have been guilty of electoral fraud. Maybe they have received donations (bribes) from vested interests. They may say one thing and do another. Exposing mistakes, corruption and hypocrisy can be effective but carries the usual risks of attacking the person and not their arguments: it can be seen as underhanded.

Usually, most of those clamouring for war are not the ones whose lives are at stake. Many of them are politicians, media commentators or public figures. A possible retort is to ask why they aren't going to the front lines or making any of the sacrifices they are expecting of others.

More generally, it is possible to question whether opinions or decisions should be made on the basis of endorsements. This is an attempt to turn the discussion

---

12 A US general often quoted for his anti-war views is Smedley D. Butler, *War Is a Racket* (Los Angeles, CA: Feral House, 2003, originally published in 1935).

from the status and prestige of people involved to a consideration of the arguments.

Challenging rewards

Questioning or opposing rewards given to war supporters is a delicate business: it can easily go wrong. Consider, for example, health and other benefits provided to veterans. Saying that these should be reduced is likely to generate hostility. More promising is to say that every injured person—whether from battle, construction work or domestic violence—should receive the same benefits and support.

Then there are the rewards for valiant acts on the battlefield, such as the Victoria Cross or Medal of Honor. For outsiders to say these are inappropriate or that they glorify killing would likely create antagonism. However, it could be effective if some of the award *recipients* question recognition of bravery.

Easiest to criticise are corporations that make huge profits from war-making. Another target is politicians who instigate or prosecute military build-ups or wars. Politicians appreciate recognition and praise for their acts; if instead they are met with protests and ridicule, they will not be pleased.

In challenging rewards, it is those whose patriotism and sacrifice are least questionable who can have the greatest impact. For example, militaristic politicians are in the best position to cut back financial benefits to veterans. In general, though, challenging rewards for those involved in war seems to be one of the least promising ways of opposing the patriotism-war connection.

***

So far, I have outlined five types of tactics for promoting patriotism in relation to war—exposure, valuing, positive interpretations, endorsement and rewards—and five corresponding counter-tactics for challenging the military-patriotism complex. Now it is time to turn to another set of tactics, involving alternatives to war. Instead of directly questioning, devaluing or confronting the system, the idea is to propose and promote a different way of doing things. An example is diplomacy. As well as saying "This war plan is foolish and likely to be disastrous" it is possible to say, "Diplomacy should be the first option."

To discuss alternatives to military preparations and war is a big task. As well as peacemaking through the efforts of professional diplomats, possibilities include reducing military expenditures, converting military production to production for civilian purposes, relying entirely on defensive-only military equipment and strategy (for example, fortifications but not tanks), using foreign aid to overcome poverty and inequality, building greater understanding of other societies (to reduce fear of foreigners) and promoting education and journalistic approaches that foster peace.

SOCIAL DEFENCE

Here, I will look at a specific alternative: defending communities through popular nonviolent action—such as rallies, strikes, boycotts and occupations—and getting rid of military defence. This is called various names: social defence, civilian-based defence, nonviolent defence and defence by civil resistance. I'll usually refer to it as social defence.[13]

Converting to social defence would involve a range of transformations. Instead of relying on troops and weapons to deter and defend against attack, people would

---

13 Anders Boserup and Andrew Mack, *War Without Weapons: Non-violence in National Defence* (London: Frances Pinter, 1974); Robert J. Burrowes, *The Strategy of Nonviolent Defense: A Gandhian Approach* (Albany, NY: State University of New York Press, 1996); Antonino Drago, *Difesa Popolare Nonviolenta: Premesse Teoriche, Principi Politici e Nuovi Scenari* (Turin: EGA, 2006); Theodor Ebert, *Gewaltfreier Aufstand: Alternative zum Bürgerkrieg [Nonviolent Insurrection: Alternative to Civil War]* (Freiburg: Rombach, 1968); Gustaaf Geeraerts (editor), *Possibilities of Civilian Defence in Western Europe* (Amsterdam: Swets and Zeitlinger, 1977); Stephen King-Hall, *Defence in the Nuclear Age* (London: Victor Gollancz, 1958); Bradford Lyttle, *National Defense Thru Nonviolent Resistance* (Chicago, IL: Shahn-ti Sena, 1958); Brian Martin, *Social Defence, Social Change* (London: Freedom Press, 1993); Johan Niezing, *Sociale Verdediging als Logisch Alternatief: Van Utopie naar Optie* (Assen, Netherlands: Van Gorcum, 1987); Michael Randle, *Civil Resistance* (London: Fontana, 1994); Adam Roberts (editor), *The Strategy of Civilian Defence: Non-violent Resistance to Aggression* (London: Faber and Faber, 1967); Gene Sharp, *Making Europe Unconquerable: The Potential of Civilian-based Deterrence and Defense* (Cambridge, MA: Ballinger, 1985); Gene Sharp with the assistance of Bruce Jenkins, *Civilian-Based Defense: A Post-Military Weapons System* (Princeton: Princeton University Press, 1990); Franklin Zahn, *Alternative to the Pentagon: Nonviolent Methods of Defending a Nation* (Nyack, NY: Fellowship Publications, 1996).

need to take responsibility for defence themselves. This would involve developing and practising skills in nonviolent action, planning for threats and contingencies, and designing technological systems so they are unattractive to enemies but instead can serve the resistance. For example, people might learn the language and culture of potential enemies, build links with opposition groups in potential aggressor states, and set up resilient communication systems.

In 1968, Soviet and other Warsaw Pact troops invaded Czechoslovakia. At that time the Soviet government dominated Eastern European countries. In Czechoslovakia, there was a reform movement in the ruling Communist Party, moderating some of the harsh controls previously imposed. This was called "socialism with a human face." These developments were threatening to the Soviet rulers, hence the invasion.

Czechoslovak military commanders decided not to resist the invasion, recognising that armed resistance would not succeed. Instead, there was a spontaneous nonviolent resistance by the Czechoslovak people, involving rallies and noncooperation.[14] The radio network broadcast messages advocating resistance and advising against any violence. The network received information that Soviet

---

14 H. Gordon Skilling, *Czechoslovakia's Interrupted Revolution* (Princeton, NJ: Princeton University Press, 1976); Joseph Wechsberg, *The Voices* (Garden City, NY: Doubleday, 1969); Philip Windsor and Adam Roberts, *Czechoslovakia 1968: Reform, Repression and Resistance* (London: Chatto and Windus, 1969).

troops were bringing jamming equipment in by rail. After broadcasting this information, workers shunted the rail car to a siding. Meanwhile, people removed street signs and house numbers so the invaders could not easily track down individuals.

Perhaps the most effective part of the resistance was talking to the invading troops and convincing them that they were doing the wrong thing. The invading Russian troops had been told they were there to stop a capitalist takeover. Czechoslovak resisters, who spoke Russian, told them "No, we support socialism, Czechoslovak-style." Many of the troops became "unreliable" and were replaced by ones who could not speak Russian.

The active phase of the resistance lasted just a week, after which Czechoslovak political leaders made unwise concessions. However, the Soviet rulers were not able to install a puppet government for eight months. The invasion and the nonviolent resistance discredited the Soviet government around the world, especially among communist parties in the west, causing many members to question Soviet leadership of the communist movement and to form independent parties. Undoubtedly the fact that resistance was nonviolent helped reduce the legitimacy of the invasion. The Czechoslovak resistance foreshadowed the Polish Solidarity movement in the 1980s and the nonviolent movements that overthrew Eastern European communist governments in 1989, including in Czechoslovakia.

The 1968 Czechoslovak resistance to the Soviet invasion was spontaneous, yet it was remarkably successful. No form of resistance had much chance of success against the overwhelming Soviet military superiority; nonviolent resistance maximised the cost to the Soviet rulers. And this was without any preparation.

Military defence is not guaranteed to be successful. Military planners recognise that to increase the prospects of success, planning, preparation and training are essential. A spontaneous armed resistance cannot be expected to succeed. The same applies to nonviolent defence: it is more likely to be effective with comprehensive training—and much else.

For example, building links with people in places where a threat might arise is valuable. In Australia, for decades some politicians and commentators drummed up a fear of an invasion from "the north"—variously Indonesia, China or Japan—used as a pretext for greater military expenditures. (In recent years, this has been superseded by alarm over terrorism.) Assuming, for the sake of argument, there was some actual threat from Indonesia (especially prior to 1998, when it was a military-based regime), social-defence preparation in Australia would involve building links with pro-democracy and anti-war groups in Indonesia. The idea is that if the Indonesian government launched an invasion, it would provide a stimulus for a challenge to the Indonesian government.

Technology is also relevant. Secure communication systems are essential to coordinate resistance and to contact allies in other parts of the world. This might involve making encryption standard, and designing systems so that no one—including the government—can monitor the content or pattern of communication. This goes right against new Australian laws that require telecommunications providers to save metadata so it can be used by security agencies in anti-terror investigations. Any system that enables centralised control is a vulnerability in the case of a foreign invasion, because it can be taken over and used by the invaders.

There is much else that could be done to build a social defence system: renewable, decentralised energy systems; factories in which workers can shut down production; resilient agricultural and transport systems.[15] Most of all, a society prepared and designed for non-violent resistance needs to be united in its goal, and in this there is a similarity with conventional patriotism. The difference is that social defence involves solidarity in defence of community, not government, and is not tied to the military.

This brings up an essential difference between social and military defence. Militaries can be used to defend against foreign enemies but are regularly used as tools by governments to defend against "internal enemies," which is code for any citizen threat to the government or the military. There are many military regimes around the world, and in most countries the military, or a militarised police, is the ultimate defender of government.

With social defence, citizens are empowered with the skills and tools to challenge repressive rulers. This means that preparations for social defence necessarily promote skills and tools that can be used to challenge the government and other powerful groups, or at least any of its

---

15 Brian Martin, *Technology for Nonviolent Struggle* (London: War Resisters' International, 2001).

policies that are unwelcome. For example, if workers have the capacity to shut down production and resist efforts to force them to get it going again—a very useful capacity in the event of a takeover—then they can use their capacity against bosses and owners. In fact, the ideal organisational form for production in a social defence system involves worker-community control, in a decentralised, cooperative arrangement. This makes it difficult for any oppressor to simply come in, replace the bosses and run the operation for their own benefit.

During the Nazi occupation of Europe, in most occupied countries the Nazis did not aim to exterminate everyone—their targets for this were Jews, Gypsies, gays and a few other groups—but rather to exploit the population and resources for their own benefit. Rather than destroy a factory, they would rather take it over and keep it operating. But the Nazi occupiers did not have the personnel and skills to replace all the managers of factories, businesses and government departments across Europe, so they relied on collaborators: citizens in the occupied countries who would serve the Nazi cause. Two prominent collaborators were Marshal Pétain in France and Vidkun Quisling in Norway; officially they were government leaders but in practice they were puppets of the Nazis. But further down the pecking order, acquiescence was also essential to Nazi rule. Business managers and government officials needed to keep doing their jobs.

In the Netherlands, there had been limited preparation in government departments for resistance to occupation.[16] Officials were supposed to do their job if it served the people but to resign if forced to implement unethical policies. However, in practice this plan was not carried out. Most Dutch government employees continued to work as usual. However, in other countries there was not even any thinking about preparing to resist.

In a social defence system, planning, preparation and training for resistance would be routine, in the same way that fire brigades plan for emergencies and run fire drills in workplaces. In the 1970s and 1980s, there was a network of a dozen social defence groups in the Netherlands, addressing different issues. One of them sought to formulate principles and plans for resistance by government employees, so they would be better prepared than they had been against the Nazis. In the 1970s and 1980s, the primary foreign threat was from the Soviet Union: there was serious concern about a Soviet invasion of Europe, and indeed the rationale for the military alliance NATO was to deter and defend against such a threat. With the end of the cold war in 1989 and the collapse of the Soviet Union in 1991, the threat evaporated and interest in social defence dissipated.

Yet the same issues remained relevant. To develop an alternative to military defence based on nonviolent resistance requires extensive planning, preparation and training. Most of all, it requires people to understand and

---

16 A. H. Heering, "Het openbaar bestuur onder vreemde besetting," *Bestuurswetenschappen*, nr 4, april/mei 1983, ("Public administration under foreign occupation," http://www.bmartin.cc/pubs/peace/83Heering.html).

be committed to unarmed resistance to aggression and oppression. This would have implications for nearly every aspect of society. The general direction for a transformation towards social defence is self-reliance, self-sufficiency, decentralised decision-making, and empowerment of citizens through skill development and training.

A society organised for social defence would be a society resistant to any form of domination—including by its own government. What this means is that if people have the understanding and skills to resist an invader, they can use the same understanding and skills to challenge the government itself, if it becomes oppressive in some way. This, in my view, is the primary reason why few governments are keen to promote social defence.

Governments are protected from internal challenges by their own systems of organised violence, primarily the military and police. In practice, most of the time these systems are not needed. Most people cooperate with laws, and support enforcement of laws. When someone steals a car or assaults a stranger, most citizens cooperate with police in tracking down the culprit. But sometimes there are serious challenges to the government or to other powerful groups, especially corporations, and so force is used to protect the system. When people refuse to pay their taxes, then the courts, and the police if necessary, are invoked to force compliance. If workers go on strike or occupy the workplace, troops are sometimes brought in to break the resistance.

Completing the picture is selective enforcement of the law: when governments break their own laws, there is seldom any penalty, and when big companies flout the law, they often get away with it or suffer only a small symbolic penalty.[17] The point is that the police and military nearly always support those with more power. Governments write laws that benefit those with power and wealth and then enforce the laws in a selective fashion, with those with little power or wealth receiving most of the blame for law-breaking.

In a society with a social defence system, ordinary members of the public would be empowered. A government that lost the trust of significant portions of the population would have a difficult time surviving. To reiterate: empowering the people to resist oppression is threatening to most governments, so social defence is unlikely to be supported. It might be okay to support people power movements in other countries, to challenge enemy regimes, but promoting equivalent movements at home is another story.

With this background, it is useful to look at tactics used by governments to oppose the option of social defence. This assessment offers some clues about how to promote this alternative.

Cover-up

Few governments give any attention to social defence. "Cover-up" is not quite the right word for this treatment, which might better be called neglect or lack of interest. The social defence option is not on the government agenda, and there are no obvious means to raise it. When

---

17 See chapter 4.

was the last time that a government sponsored a major public investigation into modes of defence?

The mass media usually follow government cues, and have given little attention to social defence. Peace movements often don't promote alternatives as much as oppose wars and weapons systems: they are better called antiwar movements.

There has been interest in social defence in a few parts of the world, including Australia, Britain, Canada and the US, but most progress in this direction occurred in Europe. This makes sense. European peoples had experience in being conquered and occupied by powerful regimes—Nazi Germany and the Soviet Union—or, if spared themselves, seeing their near neighbours being subjugated. Military defence against a much more powerful opponent was pointless or worse, except as part of an alliance with a powerful ally (the US military, via NATO). But with the collapse of the Soviet Union, much of the incentive to explore social defence evaporated. No threat, hence no need for an alternative. Of course this didn't mean governments dismantled their military systems. It meant that civil society groups became less active as the official rationale for military forces became less salient. Indeed, it might be said that governments became less active in raising alarms about invasion, and hoped that few would notice that the rationale for standing armies and advanced weapons systems was gone. Then, conveniently, terrorism apparently provided a new pretext for military preparedness. Social defence provides a template for a citizen-based alternative to conventional anti-terrorism, but this was undeveloped and never captured much interest among peace groups. After all, anti-terrorism was a pretext, and terrorism a minor problem, compared to the real possibility of nuclear attack during the cold war.

Devaluation
Governments and their apologists, on the few occasions when they took notice of social defence, could easily dismiss it as impractical—it simply wouldn't work against a determined invader. Their assumption has always been that a ruthless aggressor will always be victorious over nonviolent opposition.

This sort of dismissal by governments wouldn't matter so much except that it has long been shared by a large proportion of the population. Most people have been convinced, somewhere along the line, that violence is superior. Hollywood films assist in this: the good guys always win against bad guys by using violence, either greater force or force used in a smarter way. Few mainstream films show the power of collective nonviolent action. Despite dozens of repressive regimes having been toppled through mass citizen resistance over the past century, this has not become the stuff of Hollywood scripts. Instead, superheroes are a popular genre.

The glorification of violence as the antidote to threats to the citizenry contains an implicit devaluation of popular nonviolent action, which is assumed to be ineffectual and hence easily dismissed.

Reinterpretation
Another response to the idea of social defence is to provide arguments about why it won't work. A typical one

is to say, "It wouldn't work against the Nazis." This is less an argument than an assertion that operates by appealing to unarticulated assumptions, in particular that ruthless violence will always triumph over nonviolent action. The argument about the Nazis has been countered in several ways, for example by noting that nonviolent action was used against the Nazis in some countries, with a degree of success,[18] and more generally that nonviolent action was not even tried systematically, and certainly not as a strategy by governments.[19]

There have been few serious critiques of social defence. One of them was a study by Alex Schmid, who analysed opposition to a potential Soviet occupation of Western Europe.[20] Schmid, to his credit, also analysed armed resistance to Soviet domination, for example in Lithuania from 1944 to 1952, and found it too was ineffective. Schmid's arguments were questionable at the

---

18 Jacques Semelin, *Unarmed against Hitler: Civilian Resistance in Europe, 1939-1943* (Westport, CT: Praeger, 1993).

19 For a careful response to the argument about ruthless violence, see Ralph Summy, "Nonviolence and the case of the extremely ruthless opponent," *Pacifica Review,* Vol. 6, No. 1, 1994, pp. 1–29.

20 Alex P. Schmid, with Ellen Berends and Luuk Zonneveld, *Social Defence and Soviet Military Power: An Inquiry into the Relevance of an Alternative Defence Concept* (Leiden: Center for the Study of Social Conflict, State University of Leiden, 1985).

time.[21] Their weakness was shown more dramatically a few years later with the collapse of Eastern European communist regimes in 1989 and the dissolution of the Soviet Union in 1991, triumphs of people power against repressive regimes.[22]

Careful arguments against social defence have not played a major role in its dismissal apparently because it is easy to dismiss the option on the basis of simplistic assumptions about the superiority of violence and appeals to the Nazi example and other assumed refutations.

Official channels

Attempts to convince governments that social defence is a viable option, indeed a superior alternative to military defence, have made little progress. Gene Sharp, the world's most prominent nonviolence researcher, wrote two books about civilian-based defence and spent considerable effort seeking to convince the US government to adopt the option.[23] The US-based Civilian-Based Defense Association, which largely followed Sharp's approach, also made efforts, all to no avail. The US government never even initiated a major public investigation into civilian-based defence. Seeking change via appealing to elites turned out to be a dead end.

---

21 Brian Martin, Review of Alex P. Schmid, *Social Defence and Soviet Military Power*, in *Civilian-Based Defense: News & Opinion*, Vol. 4, No. 4, May 1988, pp. 6–11.

22 Michael Randle, *People Power: The Building of a New European Home* (Stroud, UK: Hawthorn, 1991).

23 Sharp, note 13.

A few governments have looked seriously at social defence. Sweden has a "total defence" system incorporating conventional military defence, civil defence (bomb shelters, underground factories and other preparations to survive attack), psychological defence (preparation for the possibility of war) and social defence. The idea is that in case of invasion, if military defence fails, civil defence can provide protection and the population will be psychologically prepared and able to use nonviolent means to resist. This is not the same as a social defence system, especially considering that mixing violent and nonviolent methods can undermine the effectiveness of nonviolent resistance. Still, the Swedish system nominally includes nonviolent options, though they are subordinated to conventional military means. It should be mentioned that Sweden has a well-developed arms manufacturing industry, and its arms exports are the largest in the world on a per capita basis: it is not a model for fostering nonviolent alternatives.

As mentioned, in the Netherlands in the 1970s and 1980s there was considerable grassroots interest in social defence, as well as a number of articles and books exploring and promoting this option.[24] Nevertheless, the government was not much interested, until a minor party was able to use its pivotal role to push for a dozen social defence research projects.[25] But this was reduced to a

---

24 J. P. Feddema, A. H. Heering and E. A. Huisman, *Verdediging met een Menselijk Gezicht: Grondslagen en Praktijk van Sociale Verdediging* (Amersfoort: De Horstink, 1982); Niezing, op. cit.

25 Giliam de Valk in cooperation with Johan Niezing, *Research on Civilian-Based Defence* (Amsterdam: SISWO, 1993).

single study—the Schmid study discussed earlier—which turned out to be more critical than supportive of social defence.

In Austria, conscripts are taught about social defence for part of their training. In Italy, individuals who were conscripted could opt for alternative service, and one option was being involved with an organisation promoting social defence.

Slovenia was formerly part of Yugoslavia. Around the time of the Balkan wars, Slovenia sought independence, and obtained it without any fighting. At that time, there was support for social defence. It was an optimal time for changing, especially for a small, weak state with no serious prospects of being able to defend militarily against an aggressor. But the interest in social defence faded and Slovenia ended up with a conventional military system.

The Baltic states—Latvia, Lithuania and Estonia—were independent countries when, in 1940, they were incorporated into the Soviet Union. The next year they were conquered by Nazi Germany, and then reconquered by the Soviet Union in 1944. After 1989, with the collapse of Eastern European communist regimes through mass citizen action, people in the Baltic states used nonviolent means to agitate for independence, and were successful in 1991. It was a classic case study of a nonviolent challenge to an oppressive ruler. So, some leaders thought, why not change to a social defence system and thus institutionalise this form of citizen resistance? There was interest—but only in Lithuania did interest continue. In 2015, the country's Ministry of Defence produced a manual for citizens on how to nonviolently resist an invasion.[26]

Various lessons can be drawn from these examples. One is that more pressure is needed to get governments to take social defence seriously. Another is that governments are the least likely group to make moves towards social defence. After all, if the state is built on a claimed monopoly over the legitimate use of violence in public, then social defence is a direct challenge to the state. Only the most enlightened leaders are likely to take it seriously.

Intimidation and rewards
It's possible to imagine that proponents of social defence might be subject to threats and attacks, perhaps losing their jobs or being arrested and assaulted. So far, there seems little evidence of anything like this. It would be ironic should this occur, because the methods of social defence are designed to deal with attacks.

The other side of the coin is rewards for those who support military defence, and there are plenty. Promoters and supporters can obtain careers in the military or supporting agencies, such as arms manufacturers, and bask in the recognition that comes with being part of a country's defence establishment. The entire military-industrial complex—a complex to which can be added science, education and other sectors—is built around rewards for

---

26 Maciej Bartkowski, *Nonviolent Civilian Defense to Counter Russian Hybrid Warfare* (Baltimore, MD: Johns Hopkins University Center for Advanced Governmental Studies, 2015), http://www.advanced.jhu.edu/nonviolent

those contributing. To promote social defence instead is, most likely, to forgo such rewards.

***

Social defence, as an alternative to military defence, thus faces quite a few obstacles, classified here into the categories of cover-up (though neglect is a better description), devaluation, reinterpretation, official channels and lack of rewards (whereas there are considerable rewards for supporting military defence). The next question is, how can they be countered?

Exposure
The first and essential step in promoting social defence is to make more people aware of this option. This can be done via articles, blogs, talks, debates and media coverage. This seems obvious enough. Indeed, it is far easier today to make information available than it was in the 1980s, before the Internet. Despite the apparent ease of making the concept of social defence more visible, it has not been happening. It is worth considering some factors.

One problem today is information overload. Decades ago, the main challenge was gaining access to information about social defence, which meant finding out about a newsletter, article or book and obtaining it. Today, much of the same information—in books for example—is readily available for those who want to pursue it, but it is drowned in masses of other information. This is nothing new, but the factor of overload is much more significant today.

Another problem is that information needs to be made relevant to today's circumstances. Warfare is different today than in the 1980s, and likewise social defence needs to be updated. Reading books written in the 1950s or 1980s is informative, but to engage more people in the ideas, contemporary relevance is vital. A big component of social defence today is likely to be online. Tactics, strategies, logistics and skills need updating.

Then there is the question of who is going to lead a resurgence of interest in social defence. It is all very well to talk about making the concept visible, but who will do this? In analysing tactics to promote an alternative to the war-state nexus, there need to be individuals and groups who will pursue them.

There is yet another consideration. Perhaps it is unwise to advocate directly for social defence, as this may only stimulate opposition by those committed to military defence. Another option would be to join campaigns that increase the capacity for social defence, even though that is not their purpose. Skills and strategies for overthrowing dictators are highly relevant. So are skills and strategies for challenging online surveillance, for developing local energy self-reliance, for building transport systems not dependent on imports of fuel, and a host of other areas.

Any centralised system is vulnerable to takeover. Think of transport, for example. If most people can get around by walking or cycling or vehicles powered by locally produced energy, then the transport system is resilient. Hence, the population cannot easily be subjugated by cutting off imports of oil or by occupying refineries or power plants. The same applies to communications. If a government can monitor everyone's calls and Internet usage, then the population is vulnerable to oppression by the government itself or by any aggressor that takes over the system. The implication is that efforts to build resilient transport systems and secure communication systems can make a community less vulnerable to control. This is a contribution to the capacity for social defence, even if no one ever thinks about defending nonviolently against aggression.

Social defence through changes that pass unnoticed? Is this better or worse than making more people aware of the option?

Valuing

A second aspect of promoting social defence is to increase its credibility by association with things people value. This might include endorsements by high-status people or associations with valued symbols.

Stephen King-Hall, a British naval officer in World War I, later became a prominent social commentator and an advocate of social defence. His book *Defence in the Nuclear Age,* in which he recommended abandoning military defence and defending Britain through citizen nonviolent resistance, was one of the earliest full-scale proposals for nonviolent defence.[27] For respected military personnel to give credence to social defence is a potent endorsement, because it can make people think the option is worth considering. So far, however, very few prominent

---

27 Stephen King-Hall, *Defence in the Nuclear Age* (London: Victor Gollancz, 1958).

people in any sphere of life—politicians, celebrities, business executives, religious figures, famous scientists— have endorsed social defence.

Some respected figures have endorsed nonviolence, especially those who have led campaigns: Martin Luther King, Jr, Nelson Mandela, Desmond Tutu, Aung San Suu Kyi. However, no such figure has paid much attention to social defence.

Endorsement can also come from respected organisations, but few have taken any notice of social defence, much less given it their backing. The Green Party in Germany, from its beginnings, endorsed social defence. Although green parties are often associated primarily with environmentalism—via the symbolic colour green—in principle they are built around four principles: ecological wisdom, social justice, grassroots democracy and nonviolence. However, whatever the formal policies of green parties, in practice few of them have done much to promote social defence. Perhaps this is a good thing, because it can be risky for an alternative to be identified with a political party, because then it may be more strongly opposed by members of other parties.

So far, the principal endorsements of social defence have come from those who have written about it and advocated for it. Most of those in this category have been peace researchers, such as Johan Niezing, Theodor Ebert, Gene Sharp and Johan Galtung. They add credibility to social defence in part through their status within the field, but perhaps more on the basis of what they actually write. Furthermore, most of their support for social defence was during the cold war. Johan Galtung, the world's leading peace researcher, wrote insightful essays on social defence in the 1960s,[28] but has not given the option much attention in more recent works. Gene Sharp, the world's most prominent analyst of nonviolent action, wrote two important books about civilian-based defence in the decade before the end of the cold war. Since then, Sharp has received quite a bit of mainstream recognition for his work on nonviolent action, especially in the wake of the Arab spring, but this has not had much spin-off for civilian-based defence.

In summary, social defence has received few endorsements outside of small community of scholars and activists who study and support it. This no doubt has contributed to its marginalisation.

Interpretation
Social defence, when it is raised with audiences unfamiliar with it, receives a variety of responses. Some people dismiss it out of hand; a few are intrigued and want to know more. However, these responses are mostly at the gut level, based on emotions and assumptions. At the intellectual or cognitive level, though, there can be a calm, logical engagement with arguments and evidence. At this level, advocates of social defence can make quite a few points.

---

28 Johan Galtung, *Peace, War and Defense: Essays in Peace Research, Volume Two* (Copenhagen: Christian Ejlers, 1976), pp. 305–426.

• Military defence cannot easily be separated from military offence: systems nominally set up for defence can be used for aggressive or interventionist purposes.

• Arms manufacture and sales underlie a huge amount of killing and suffering throughout the world.

• Military forces, in many countries, are used to support authoritarian governments.

• Social defence is based on methods of nonviolent action that have been shown to be more effective than armed struggle against repressive governments.

• Social defence is a system in which the means reflect the ends: if the goal is a world in which conflict is carried out without violence, then it is desirable that the methods to achieve such a world should not involve violence. (In contrast, military systems use the threat of violence to pursue "peace.")

• Social defence can build a sense of solidarity among people, because preparations require this.

• Social defence systems promote skills throughout the population, including skills in persuasion, communication, decision-making, protest, noncooperation, and self-reliance in energy, transportation, agriculture and other arenas.

• People who learn the skills for social defence can use those same skills to pursue social justice, for example to challenge government repression and corporate abuses.

However, such arguments are unlikely to win over anyone who is not already sympathetic.

Mobilisation of support

Gene Sharp, who wrote important books about civilian-based defence, believed that governments could be convinced to switch to this alternative after they were shown it was more effective, but his efforts were unsuccessful. Indeed, although he received some polite hearings, the US government made no significant initiatives towards civilian-based defence—not even an official investigation—meanwhile spending hundreds of billions of dollars every year on the military. This is a telling example of how logic and evidence cannot make much headway in the face of deeply held beliefs linked to vested interests. It might also indicate that the real driving force behind US military preparedness is not defence against foreign enemies but rather protection of US state and corporate interests.

Trying to convince government and military leaders about the effectiveness of social defence is to use official channels to bring about change. This is unlikely to be successful, and indeed official channels such as government inquiries or expert panels often serve to give the appearance of dealing with concerns while actually nothing much happens. My view is that governments are the least likely to take the initiative to introduce social defence, because they have the strongest stake in having military forces to protect their own interests.

Instead of appealing to governments, the alternative is to mobilise support. For promoting social defence, this means building popular support via a mass movement, in the spirit of previous movements: anti-slavery, labour, feminist, peace, environmental, animal rights and other movements. A movement for social defence could start out as a subset of the peace movement, but to have any chance of success it needs to have a wider base. The labour movement is important because, in a social defence system, workers need to be prepared and skilled in withdrawing and/or using their labour to resist impositions by an aggressor. Social defence is also relevant to most other movements, via the skills needed for resistance and via reorganisation of society to have the solidarity to oppose aggression and repression.

In relation to patriotism, there is a complication. Civilian-based defence, as presented by Sharp and others, is seen as national defence, namely defence against foreign aggressors. The idea is to replace one form of national defence by another: military defence becomes nonviolent defence. Much of the advocacy for civilian-based defence is built around this assumption. This has the advantage of conforming to the usual thinking about defence, and drawing on assumptions about nationalism and patriotism. It does not question conventional government-promoted views about the military and its purposes.

Treating civilian-based defence as national defence is at the same time a disadvantage. It assumes that state and military leaders are the ones who will make decisions to switch to a different form of defence, when they are the least likely to want to make such a change.

Another way to think of social defence is as defence of a community by its members. The word "community" is vague and makes assumptions about relationships between individuals. The idea, though, is that the state or nation is not necessarily the unit being defended. A more likely possibility is that people defend themselves against their own government, including against troops or militarised police. "Social defence" in this formulation is defence against government repression. This is actually the usual meaning of social defence in some European countries. It makes sense in relation to the dual purpose of military forces: to defend the state against external and internal enemies. The internal "enemies," in many cases, are simply citizens who are challenging abuse of power by the government. This is another way of seeing why few government leaders are likely to be convinced to switch from military to social defence.

Mobilisation of support for social defence means getting individuals and groups to support and take action to strengthen people's commitment and skills to resist aggression and repression and to develop plans and build infrastructure to enable this. Since the 1990s, only a few groups in a few countries have been advocating for social defence, so most of the progress is happening in indirect ways.

• The spreading of skills in nonviolent action against repressive governments. This is ideal preparation for social defence. In fact, people power movements are social defence in action. What they lack is any sustained way of creating a system for nonviolent resistance as an alternative to military defence.

• Network communication systems, using phones, texts, Facebook, Twitter and other social media. Repressive governments can more easily control one-directional media such as television and newspapers; networked media are more readily used for resistance. However, governments are increasingly collecting data from social media to monitor dissent, so methods of opposing surveillance, such as encryption, are important to enable resistance.

• Technological self-reliance. Movements for local food production, decentralised energy production, and transport by walking and cycling help to make local communities less dependent on centralised facilities that can be controlled by governments.

• Protest movements—against poverty, exploitation and a host of other injustices—can provide experience and understanding in how to oppose repression, especially when the movements involve mass participation using methods of nonviolent action.

These and other developments are building capacity that can be used against foreign aggressors and against home-grown repressive governments. Whether this is an adequate substitute for a social defence system is another matter. Almost certainly it is not.

Governments continue to develop their capacities to control their own populations, for example through monitoring of dissent through mass surveillance and targeted intelligence operations, sophisticated public relations operations, suppression or cooption of initiatives for worker self-management and participatory democracy, and promotion of high-tech infrastructure—large power plants, industrial agriculture dependent on pesticides, high-rise buildings—that is high cost, potentially vulnerable to disruption and amenable to centralised control. In the context of defending against aggression, campaigns against this type of infrastructure contribute to making communities less vulnerable to attack and domination.

Resistance to intimidation and rewards

Supporters of military systems, to oppose critics and challengers, can intimidate them and/or offer rewards to tempt them to change their views or actions. A typical sort of intimidation is the surveillance, infiltration, disruption and repression of peace groups. Typical rewards include jobs and funding for supporters of the military, including individuals, companies and sectors of the population. These methods are likely to be used against promotion of social defence, at least if this promotion gains traction.

Promoters of social defence therefore need to be prepared to resist intimidation. This is a perfect example of methods reflecting and serving goals: the goal is a system for citizens to nonviolently defend against aggression and repression, and to promote this goal it may be necessary to defend against repression. At the moment, advocacy for social defence scarcely exists, and the risk of repression is not so great. It can be expected that if a significant movement develops and starts making progress promoting and implementing social defence, elements within the military may take serious steps to subvert or crush the movement.

Countering rewards often can be more difficult than countering intimidation. There are vastly more research grants and career opportunities for military-related projects than ones involving nonviolent action. Promising nonviolence practitioners and researchers may be attracted to jobs in the system that seem worthwhile but restrain activism. Resisting temptations is part of promoting alternatives to the military. The bigger task is to change the incentive structure. This is a huge challenge. Imagine the hundreds of billions of dollars now spent on military systems every year being redirected to the building and maintenance of social defence systems. This would indeed be a revolution in defence affairs.

Conclusion

There are two main ways to challenge state-centred thinking linked to military systems. One is to directly respond to the war machine, addressing the massive attention to war, the glorification of military sacrifice, the rationales for military forces, the institutional legitimation of "defence," and the intimidation of critics. Antiwar movements have made an enormous difference in deterring or helping halt particular wars and opposing particular weapons systems. Even so, the war system remains central to the world order, because military forces serve a dual role, protecting the state against both external enemies and internal challenges.

A second way to challenge military nationalism is to propose alternatives to military defence. I examined one particular alternative, social defence, that involves preparations for citizens to resist aggression and repression, through understanding, training and choice of appropriate technological systems. This option has been almost completely marginalised. Nevertheless, an analysis of tactics can be helpful in seeing ways to promote social defence and the barriers likely to be encountered. In order to be a challenge to state-based defence, social defence needs to be conceptualised as community defence, in many cases against the state. This potential for undermining state power is probably a primary reason why few governments have made any steps towards converting from military to social defence, or even investigating the possibility.

14
Investigating tactics

The world seems to be made up of countries, and each one has a government. Most people think about the world in terms of countries and governments—it seems natural, and it's a convenient way to make sense of the news and much everyday discussion. In addition, many people identify with a particular country.

This way of seeing the world, while useful for some purposes, can be misleading. Countries and governments are not the same as the people in them. Furthermore, governments often act contrary to the interests of the population, instead serving the interests of those with the most wealth and power.

What I call "ruling tactics," which might also be called "patriotism tactics," are methods used to encourage people to think in terms of countries and to identify with a particular one, and not to question in any fundamental way how wealth and power are distributed. To illustrate these tactics, I've chosen a variety of issues such as sport and terrorism. I picked these particular examples because I know something about them and they are addressed in everyday conversations. However, my assessments are far from definitive. Much more could be done to examine tactics in relation to any given topic, and to tackle additional topics. If you want to do this, how should you proceed?

Consider a topic such as the economy, transport or the arts. Concerning this topic, a useful first step is to ask whether there is any plausible reason for people to identify with a country. (See chapter 2.) Does the average person really have much in common with thousands or millions of others who will never be friends or even be introduced, just because they are resident in the same area of land? For example, it might sound beneficial that the economy has grown, but looking more closely it could be that nearly all the increased income has gone to the top 1% of earners: not everyone has the same stake in *our* economy, namely the country's economy. If you're one of the wealthy ones, fine, but otherwise thinking in terms of *our* economy is misleading. It's even more misleading if you take into account people in Bangladesh, Malawi and Peru.

The next step is to look at the common types of tactics used by rulers and their supporters to gain support for the system. As listed in the introduction, five tactics are commonly used in relation to the system (the country, the nation, the government).

System-support tactics
1. Exposure (of positives); attention
2. Valuing
3. Positive interpretation
4. Endorsement
5. Rewards

So you look for evidence of any of these tactics, for example in the media, government policies or everyday conversations. For example, what sorts of comments are there are about economic growth, especially of *our* economy? If growth is mentioned, that's exposure. If it's seen as a good thing, that's valuing—and so forth.

After looking at tactics to encourage support for the system, you can look at tactics to oppose challenges and alternatives. To do this, you might need to learn more about a particular alternative or campaign, such as the Occupy movement, the global justice movement or the steady-state economy. Then you look for evidence about tactics against this alternative or campaign. The five tactics listed in chapter 1, regularly used against challenges to the dominant view, are

System-support tactics: opposing challenges and alternatives
1. Cover-up
2. Devaluing
3. Negative interpretation
4. Discrediting endorsements
5. Intimidation

For example, you might notice that there is little or no discussion of steady-state economics—it is covered up— or that when it is discussed, it is criticised or dismissed as irrelevant or foolish.

Some of these techniques are more visible than others, depending on the topic. Some alternatives are hardly ever discussed: cover-up is so effective that other techniques are not required. In relation to conventional economics, Gandhian economics is one such alternative. However, some economic alternatives occasionally obtain visibility, for example local currencies, the Tobin tax or a guaranteed annual income, in which case you need to examine the way they are treated by various commentators and spokespeople. Evidence of devaluation and reinterpretation can come from what people say and write. Evidence of intimidation can sometimes be hard to obtain: it is hidden. Local currencies have sometimes been shut down by governments, but this is not widely known.

If you're involved in a campaign to challenge dominant perspectives and promote alternatives, then you can go on to challenger tactics.

Opposing system-support tactics
1. Exposure (of negatives)
2. Devaluing
3. Negative interpretation
4. Discrediting endorsements
5. Refusing rewards

Promoting alternatives
1. Exposure
2. Valuing
3. Positive interpretation
4. Endorsement
5. Rewards

These tactics provide a rough framework for thinking through how to proceed and in particular for seeing whether there are actions that might be taken. For example, if you are involved in promoting local currencies, you can oppose the tactics by rulers by exposing the negative consequences of the conventional money system, devaluing it, explaining what is wrong with it, and so forth. You can promote the alternative by publicising and valuing it, and so on.

These are big topics, and no one can do everything. To become deeply involved in just one issue such as local currencies can become a life's work and, depending on the individual, it can be worthwhile putting most effort into one or two tactics, for example explaining the alternative to wider audiences or trying to implement it in a particular area.

Another possibility is to look at what's being done already and seeing whether there are any significant gaps, namely worthwhile tasks that are being neglected. This could be an opportunity to make a difference.

It is helpful to remember that countries, borders and states are human creations. They are all fairly new, and are neither inevitable nor necessary aspects of the way humans organise themselves. The fact that so many people spend so much effort encouraging everyone to think in terms of countries and governments indicates that this perspective does not come naturally. There are strong contrary pressures to think locally and globally. System-support tactics are just tactics, not guaranteed to succeed. Understanding them makes it easier to resist them more effectively.

Index

www.ingramcontent.com/pod-product-compliance
Lightning Source LLC
Chambersburg PA
CBHW060344200326
41519CB00011BA/2037